Helion & Company Limited
Unit 8 Amherst Business Centre
Budbrooke Road
Warwick
CV34 5WE
England
Tel. 01926 499 619
Email: info@helion.co.uk
Website: www.helion.co.uk
Twitter: @helionbooks
Visit our blog https://helionbooks.wordpress.com/

Typeset by Mach 3 Solutions (www.mach3solutions.co.uk)
Cover design by Paul Hewitt, Battlefield Design (www.battlefield-design.co.uk)

ISBN 978-1-804513-74-3

British Library Cataloguing-in-Publication Data
A catalogue record for this book is available from the British Library

We always welcome receiving book proposals from prospective authors.

CONTENTS

To Suzanne and Arno

Note: In order to simplify the use of this book, all names, locations and geographic designations are as provided in *The Times World Atlas*, or other traditionally accepted major sources of reference, as of the time of described events.

ABBREVIATIONS AND ACRONYMS

AAM	air-to-air missile
BAF	Bangladesh Air Force (*Bangladesh Biman Bahini*)
CATIC	China National Aero-Technology Import and Export Corporation
CKD	component knock-down (kits)
DME	distance measuring equipment
DPRK	Democratic People's Republic of Korea (North Korea)
g	gravities
GPS	Global Positioning System
HOTAS	Hands-on Throttle and Stick
HUD	Heads-up Display
HUDWAC	Heads-up Display Weapon Aiming Computer
IFF	Identification Friend-or-Foe
ILS	instrument landing system
IRIAF	Islamic Republic of Iran Air Force
kg	kilogramme
km/h	kilometres per hour
m/s	metres per second
PAF	Pakistan Air Force
PLAAF	People's Liberation Army Air Force (Chinese air force)
PLANAF	People's Liberation Army Navy Air Force (Chinese navy aviation)
PRC	People's Republic of China
RWR	radar warning receiver
TACAN	tactical air navigation
TFB	tactical fighter base
USA	United States of America
USSR	Union of Soviet Socialist Republics
VOR	VHF omnidirectional range (beacon)

ACKNOWLEDGEMENTS

The author's special thanks go to Dietrich Banach for providing his manuscript, Jürgen Vogt for numerous comments, corrections and additions, Guido E. Bühlmann for his unique photographic material and Stefan Büttner and Alexander Golz for their magnificent photos, as well as for their many years of travelling with me.

The author would also like to thank all those who have contributed in any way to the creation of this book – be it through contributions, pictures, texts, books or by opening doors:

Wing Commander Asad, Air Marshal Muhammad Enamul Bari, Winston Brent, Victor Bertschi, dlsunshy, Dr Jakub Fojtik, Siegmar Frenzel, Marcus Fülber, Air Vice Marshal M. Naim Hassan, Group Captain M.D. Shaharul Huda, Wing Commander Mahmud Mehedi Hussain, Group Captain M. Saeedul Hassan Khan, Awais Lali, Lieutenant Colonel (GS) Holger Leukert, Chris Lofting, Ian Malcom, Group Captain Abidemi Timothy Marquis, Matthias Muth, Air Commodore M. Mafidur Rahman, Air Commodore Mohammad Obaidur Rahman, Mohammad Razzazan, Andreas Rupprecht, Air Commodore Sarfraz Ahmed Khan, Ulrich Schneider, Lieutenant Colonel (GS) Carlo G. Schnell, Erik Sleutelberg, Kai Spörhase, Lukáš Syrovy, Babak Taghvaee, Dr István Toperczer, Jiří Vaněk, Marc van Zon.

INTRODUCTION

The development of the Chinese MiG-21 derivatives is not only an interesting technological history in itself, but also reflects the changes in the politics of the People's Republic of China (PRC) on a smaller scale: from its alliance with the Union of Soviet Socialist Republics (USSR) to rapprochement with the West and the PRC's own path of state capitalism.

Starting out as a licence-built version of the USSR's most modern tactical fighter aircraft at the time, the J-7 was continuously developed by the Chinese aviation industry over five decades as it moved from its Soviet-influenced roots to independent structures. While the evolution of the Soviet MiG-21 was complete 15 years after its first flight, the independent Chinese further development of the type was only just gaining momentum at that time. When the MiG-29, the next-but-one model, was already in service in the USSR, the first extensively modified versions of this type were just reaching production readiness in China.

In retrospect, it is clear what a stroke of luck it was for the development of the Chinese aviation industry that the MiG-21 was available to it as a model, initially for reproduction and later for further development. The comparatively simple technology of the MiG-21 represented the maximum that Chinese industry could cope with technologically at that time. The long time it took to acquire this technology makes this clear. At the same time, however, the MiG-21's flight performance was in line with the state of international technical development. In view of the protracted and setback-ridden development of the Chinese MiG-21, all Chinese indigenous developments at that time – especially if they had features that went beyond the technical level of the MiG-21 – can only be regarded as highly unrealistic and based on wishful thinking. The numerous what-if discussions in the relevant literature and in the social media therefore lack any basis.

Unlike the Soviet MiG-21, where new versions were always delivered for domestic use first, the development of the Chinese MiG-21 was very much driven by exports. Export customers often received important innovations before China's own air forces, and the main customer, Pakistan in particular, had both driven the development with requirements and supported it with experience and access to Western technology. Because there was a lack of suitable indigenous equipment components for the exported fighter aircraft in particular, the corresponding technology was purchased from Western countries. The result was versions of the F-7 that were technically superior to the J-7 for domestic use. And so it was with the J-7 that the Chinese

The author (right) visiting the Bangladesh Air Force in 2014. (Stefan Büttner)

aviation industry acquired the ability to integrate, replicate and further develop Western avionics and weapons systems. With the evolution of the J-7, it thus created the basis for today's indigenous developments to the current state-of-the-art.

The fact that the MiG-21 had become an icon of the Cold War, the most widely used and, above all, the longest-produced fighter aircraft of all time was due, not least, to its Chinese offshoot, the J-7, and its export variant, the F-7. A third of the approximately 50 MiG-21 users worldwide deployed and continued to deploy Chinese MiG-21 versions, half of which (Albania, Iran, Myanmar, Namibia, Pakistan, Zimbabwe and Sri Lanka) had only become operators of this aircraft type thanks to Chinese deliveries. In Bangladesh, Nigeria and Tanzania, Chinese F-7s replaced the Soviet-built MiG-21s previously deployed there. Several user countries – Egypt, Iraq, North Korea (DPRK), the USA and China itself – operated MiG-21s and J/F-7s simultaneously. And while Soviet series production of the MiG-21 was discontinued in 1986 after 27 years, Chinese production was only really taking-off and ended after a further 30 years in September 2016.

While the author had access to original documents such as maintenance manuals, logbooks etc. during his (short-lived) work on the MiG-21 and also afterwards, and could read Soviet documentation – with some difficulty – in the original Russian language, or at least verify facts from it, the situation was different in the case of the J/F-7: here he could not rely on original documents and his lack of knowledge of the Chinese language considerably restricted access to native sources, already hardly accessible on their own. The research for this book was therefore limited to machine-translated texts with all their limitations and to the sparse English-language secondary literature. For example, a comprehensive account of the operational history of only one user air force – the Pakistan Air Force (PAF) – was available. Nevertheless, in seven user countries – Egypt, Bangladesh, China, Iran, Nigeria, Pakistan and Sri Lanka – the author was able to inspect J/F-7s, mostly in action, and get close to the aircraft in various ways. In view of the lack of reliable information on the development process, the focus of the following descriptions was on its results, i.e. the individual versions of the J/F-7 and their obvious characteristics and capabilities. In addition, the presentation was explicitly from the point of view of those who were familiar with Soviet technology.

The keyword 'sources' is linked to another sensitive topic: copyright. The research for this book quickly made it clear that the impression that copyright is not taken very seriously in China was not entirely wrong. In just the Chinese publication *Aviation Archives 03/2008* – the primary source for this book – six of this author's images, which were available in high resolution on his website www.mig-21.de for some time, were used without reference to the source; in the online version there were seven images, some with different captions. Numerous others were circulating on the World Wide Web, also without reference to the source. The author has endeavoured to avoid this practice by using his own images and those expressly made available to him by the photographers. Nevertheless, for the sake of completeness, it was unavoidable to use images whose origin could not be conclusively clarified.

While the development history of the Soviet MiG-21 could be presented as a chronological sequence of technical development steps – the social conditions relevant to the creation and further development of the type remained largely unchanged during this time – the development of the J-7 cannot be told without looking at the enormous – and often terrible – upheavals in China at the time.

The author together with FT-7 technicians of the Nigerian Air Force at the NAF Air Expo 2010. (Holger Müller)

In order to provide a complete picture, the development of all important J/F-7 versions for domestic and export use is described below. However, a detailed description of the use of the former is reserved for future publications.

Some notes on the terminology: The Chinese MiG-21 versions exclusively for domestic use are always referred to below as J-7, those for export as F-7. When referring to 'China' and 'Chinese' in the following, the People's Republic of China is always meant. The Republic of China is referred to as 'Taiwan'. In view of the sometimes confusing and frequently changing names of Chinese authorities and state institutions, these are named according to their function, even if the official name is at least occasionally different. For example, in the following, the term 'Ministry of Aviation Industry' is usually used instead of 'Third Ministry of Machinery'.

While today's Chinese aircraft developments take place in the public eye, so to speak, thanks to the omnipresence of social media (at least as far as the visible results in the form of flying prototypes and production models are concerned), the development of the J/F-7 still took place largely covertly 'behind the Chinese wall'. Many details were therefore hidden. This book is therefore expressly not intended as the last word on the subject, but as a basis for discussion to gain further insights. The author welcomes reactions from his readers – comments, additions and questions. You can reach the author at holger.mueller@mig-21.de.

The Difficult Search for Information

It was clear at the start of work on this book that it would be difficult to gain reliable information due to the PRC's restrictive information policy. Detailed technical information was therefore to be obtained through visits to various Chinese aviation museums. The PRC's increased lockdown as a result of the corona pandemic and the treatment of foreign visitors to the country, which can only be described as 'hostile', cancelled these plans.

A second approach was based on the realisation that essential parts of the development of the J/F-7 would not have been feasible, or not in this form, without Western support. It should be possible to find out technical details from the Western partner companies, especially as the relevant cooperation took place decades ago.

The first port of call was the Rochester Aviation Archives, a collection of avionics components, brochures, company newspapers, films, videos and photographs from the companies, Elliott Bros, GEC-Marconi and BAE Systems at BAE's site in Rochester, UK. As can be seen on the website of the archive https://rochesteravionicarchives.co.uk, copies of the components developed for China (head-up display and weapon computer) are also stored there. When asked about this, the author was informed that BAE continues to classify all information on the subject as sensitive and therefore no support in the form of images or information is possible. Interestingly, following the author's enquiry, all references to the F-7 on the Rochester Aviation Archives website were removed and replaced with the more innocuous 'Project 956'.

Martin-Baker, the supplier of the ejection seats for the Pakistani and parts of the Bangladeshi F-7, responded to the author's request for information with a counter-question about the budget available for the research. After the author was only able to offer the prospect of a symbolic contribution due to his non-commercial involvement with the subject, Martin-Baker failed to respond further.

There was no reaction at all from Leonardo, the legal successor to FIAR (Fabbrica Italiana Apparecchiature Radioelettriche), the supplier of the radars for the Pakistani F-7s, or from the US Air Force, which procured F-7s from China for its Constant Peg training programme.

The author would therefore be particularly pleased to receive reactions from insiders who can contribute more on the topics mentioned from their own experience.

One possible reason for this behaviour was the realisation that China's path from developing nation to world power was only possible with Western support. It is conceivable that at a time when China was regarded by the West as a counterweight to the USSR and strengthened accordingly, more Western expertise flowed into China, particularly in the field of armaments, than would have been desirable from today's perspective. Similar to the Islamists in Afghanistan and in the Arab world, who were seen by the West and above all the USA as 'the enemy of my enemy' and supported as quasi-allies, who later turned against their sponsors, China has turned from partner to adversary.

1
EARLY HISTORY

When it was founded in 1949, the People's Republic of China was a backward agrarian state with no industrial base to speak of and no defence industry of its own. From the outset, state founder Mao therefore endeavoured to develop a modern armaments industry of his own, as he believed that only a self-sufficient state in this respect could gain weight on the international stage. Initially, the focus was on heavy conventional armaments for all three branches of the armed forces as China initially only had production capacity for light weapons, and with a later focus also on missiles and nuclear weapons. For ideological reasons, only the USSR was available as a partner for this. On the one hand, the USSR was interested in strengthening China as part of the socialist camp, but feared – not without good reason, as the Sino-Soviet border conflict in 1969 would show – that the recipient could also use the newly acquired weapons against its northern neighbour.

Immediately after the founding of the People's Republic of China, preparations for the development of a domestic aviation industry began with Soviet support. In January 1951, a Chinese delegation travelled to the USSR to negotiate support for this development and in October of the same year, the two countries signed an agreement on technical support for the establishment of aircraft repair plants. According to Chinese plans, these factories were to be converted into aircraft production plants and be able to deliver 3,600 aircraft per year.

Initially, the establishment of trial production of the Yak-18 and MiG-15 was discussed, but it was ultimately decided to produce the Yak-18 as the CJ-5 and the MiG-15's successor, the MiG-17F, as the J-5 as part of the first five-year plan. The engines for both aircraft were also to be produced in the country. The first CJ-5 then took-off on its maiden flight on 11 July 1954, the J-5 on 19 July 1956. These rapid successes in a country with little industrial tradition and a weak economic base were of course only possible thanks to massive Soviet support in terms of people and materials.

As part of a series of agreements between the two countries, the USSR sent a total of 856 specialists and managers to the Chinese aviation industry. In the following years, the USSR provided China with a total of seven types of aircraft and nine types of engines, several models of tactical missiles and hundreds of technological processes for the manufacture of on-board equipment, and supported the production of complete sets of equipment.

In May 1953, China and the USSR signed a contract initially for 143, and later 156, industrial projects, 13 of which were to manufacture aviation products such as aircraft, engines and equipment. The construction of the plants began immediately afterwards. The first production facilities for jet aircraft and their engines were the aircraft factory and the engine factory in Shenyang. At the same time, China began to build up domestic capacity for the production of aviation materials as the basis for aircraft construction. The production of important aircraft components such as rubber products and tyres, bearings and pipes, plastics, paints and textiles, fuels and lubricants was set up. According to Chinese figures, 81 percent and 78 percent of the raw materials used in the series production of the J-5 and its WP-5 engine were domestic. Almost all of the equipment was sourced from the USSR. In addition, the Eastern superpower granted corresponding loans among others over 520 million roubles during Khrushchev's visit to China on the occasion of the fifth anniversary of the People's Republic on 1 October 1954.

The Soviet side initially remained reluctant to supply nuclear technology, but in March 1955, under the impression of nuclear threats from the USA against China, which in turn threatened Taiwan, it agreed to a transfer of nuclear weapon technology. In October 1957, an agreement was then concluded on the delivery of a prototype nuclear bomb. The Chinese were also to receive the know-how for the development of launchers and so Soviet nuclear weapons and missile specialists were sent to the neighbouring country to support the development work. In addition, the Chinese pressed for the transfer of knowledge for the construction of nuclear submarines, which the USSR initially did not agree to. However, the Soviets agreed to supply further technology in a treaty dated 4 February 1959.

No sooner had the PRC's industrialisation efforts borne their first fruits than the Communist Party leadership under Mao called for a 'march towards modern science', with the aim of raising the People's Republic to the level of the leading industrial nations within 12 years. Based on Soviet supplies, this industrialisation programme (de facto a militarisation programme) was to turn China into a major economic and military power. These claims were being made in disregard of fundamental economic and technical laws, which would lead to enormous upheavals. According to official statistics, during the first five-year plan (1953–1957) 61 percent of the budget was used for the military and armaments. Major investments were made in basic research and structures were created in the form of specialised research facilities. In addition, the Soviet models were not only to be copied, but constructive Chinese ideas were to be introduced from the outset. The indigenous development of weapons systems was accelerated, even though their technological requirements were in stark contrast to the technical and technological level of Chinese industry at the time. At the same time, responsibilities were relaxed, the authority of factory management was undermined and experiments replaced systematic procedures. Above all, this caused massive quality problems, with the result that large parts of the production could not be delivered or could only be delivered after costly reworking.

All these measures culminated in the 'more and faster' of the 'great leap forward', which ultimately represented the ideological formulation of the deliberate negation of economic and technical laws. Although the 'revolutionary' measures – such as the construction of thousands of small blast furnaces – helped to increase production figures in the short term, they did so at the expense of quality and sustainability, which ultimately also affected the production of defence equipment and meant that numerous Chinese-made weapons systems were simply unusable. As a result, it was not only the development of the industry that suffered massive setbacks, rather, the 'Great Leap' caused what is probably the greatest humanitarian catastrophe in the history of mankind.

Chinese exports were massively increased to finance investments in major industrial projects. Due to a lack of competitive industrial goods, these mainly comprised agricultural products, which, together with the already existing scarcity of arable land and numerous political measures, contributed to a massive famine that claimed the lives of an incredible 30 to 45 million people (depending on the source) between 1959 and 1961.

It is a testament to the will and resilience of the Chinese workers and engineers that further successes could be achieved on the way to a domestic defence industry during this time. In 1958, a MiG-

With the licence production of the MiG-17 (as the J-5), China gained its first experience in the construction of jet aircraft. In the course of further development, the JJ-5 two-seater came into being: a version that did not exist in the country of origin. (Alexander Golz)

19 made from Soviet parts took-off on its maiden flight as the J-6, followed the subsequent year by the Tu-16 as the H-6. At the same time, the trial production of missiles, including air-to-air missiles (AAM), began.

After the Great Leap Forward had brought chaos and suffering to China for three years, moderate forces in the country's leadership prevailed and a phase of consolidation began in 1960. The uncontrolled growth of new projects was ended, the organisation of operational processes was restored and measures were taken to ensure production quality. Products of dubious quality that had been manufactured during the years of the Great Leap were reworked. In 1960, the China Aeronautical Administration (also known as the Sixth Institution) was established as the central research institution of the aviation industry. In the work of the aviation industry, the regulations that had been suspended during the Great Leap Forward were reinstated. A systematic, step-by-step approach to research, development and production became the rule again.

As a result, the Chinese aviation industry achieved numerous milestones. Production of the first PL-1 AAM (licence-built from the Soviet RS-2US) began. The J-6 and the Z-5 (Mi-4) helicopter were transferred to series production. Between 1963 and 1965, 1,055 aeroplanes and helicopters left the factory halls – a remarkable result for a country that 15 years earlier had no aviation industry to speak of.

The Rift

The estrangement between the USSR and the PRC began as early as Khrushchev's secret speech 'On the Cult of Personality and its Consequences' at the 20th Party Congress of the Communist Parry of the USSR in 1956, when the General Secretary of the Party criticised, among other things, the mass murders and forced collectivisation initiated by Stalin, measures that were commonplace in Mao's China at the time. Above all, the PRC's behaviour towards the USSR raised doubts. This included the Chinese artillery shelling of the island of Kinmen, which belonged to Taiwan, in August 1958, which took place without prior notification of the USSR and provoked the intervention of the USA, which ran counter to the Soviet policy of détente at the time. At the same time, the Chinese deliberately delayed the handover of a US-made GAR-2/AIM-9B Sidewinder air-to-air missile, which they had captured in aerial combat over the Taiwan Strait. Only after the USSR withheld design documents for the R-12 medium-range missile the Sidewinder was handed over – albeit without the crucial infrared seeker, much to the annoyance of the Soviets. As a result of this, and worried about the chaos in China after the Great Leap Forward, the transfer of nuclear know-how agreed in October 1957 was initially delayed by the Soviets and then stopped altogether on 20 June 1959. At the same time, the Chinese leadership complained about a lack of support from the USSR in the burgeoning conflict with India over the border region in the Himalayas, which ultimately led to a border war in 1962.

In June 1960, the growing estrangement became public when Khrushchev and Peng Zhen, a member of the Politburo of the Chinese Communist Party, clashed publicly at a congress of the Romanian Communist Party in Bucharest. Khrushchev's reaction did not fail to materialise. After the Bucharest conference, the USSR began to exert political, economic and military pressure on China and massively reduced economic and technical aid.

In July 1960, following previous provocations by the Chinese, the USSR decided to recall all experts working in China; over 1,000 in total, 143 of them in the aviation industry. By the end of August, all Soviet consultants working in the aviation industry had left the country. Their departure had a massive impact on development and production in the 22 factories and research institutes affected in the aviation industry. In addition to the withdrawal of experts, all deliveries of original drawings, technical information, finished products and important raw materials for aircraft were cancelled. According to Chinese figures, the economic losses for the aviation industry amounted to 546 million yuan. With the loss of Soviet support and the simultaneous technical and economic embargo by Western countries, numerous Chinese technology projects were placed in jeopardy, including the planned licensed production of the MiG-21. Licensed production of the MiG-19 and Mi-4 had been brought to a standstill. The indigenous developments, Dongfeng 107 and Dongfeng 113, were being cancelled and there was no sign of a future model to equip the air force.

In view of the massive consequences of the ideological disputes, the Chinese rowed back and showed themselves to be conciliatory at a meeting of communist parties in November of that year. The USSR then agreed to the continuation of a good third of the projects. Strategic weapons systems, such as nuclear weapons and their

While the MiG-19 was only a temporary solution in the Soviet Union until the Mach 2-capable MiG-21 was available, in China it became the standard model for three decades and the most-built fighter aircraft - not least because the industry, weakened by the Great Leap Forward and the Cultural Revolution, was initially overwhelmed by the MiG-21. The production output – allegedly at around 4,500 examples, including the A-5 derivative – almost doubled that of the USSR. (Alexander Golz)

delivery systems, were no longer transferred – but licences, pattern aircraft and raw materials for the production of the MiG-21 in China were.

The MiG-21 was the last MiG fighter aircraft introduced by China. Shortly afterwards, Sino-Soviet relations hit rock bottom again and the decade-long Soviet support for Chinese aviation came to an end. Only 30 years later, after the once mighty USSR had dissolved and Russia had taken its place, did China again have the opportunity to acquire production licences for modern fighter aircraft – but then Sukhoi took the place of Mikoyan-Gurevich.

2
FROM SOVIET TO CHINESE FIGHTER AIRCRAFT

Since the early 1950s, the Soviet Experimental Design Bureau No. 155 Mikoyan-Gurevich (OKB MiG) had been working on the development of a Mach 2-capable successor to its supersonic MiG-19, which itself was only just entering series production, based on the preliminary work of the Central Aero and Hydrodynamic Institute (ZAGI). Ever since the prototypes of the new model, which was to go into series production as the MiG-21, appeared at the air parade in Tushino in July 1956, the Chinese leadership had been aware of the existence of this model and, as with its predecessor, Beijing was very interested in setting up licence production.

On 7 February 1959, China's Prime Minister Zhou Enlai and Anastas Mikoyan, Chairman of the Soviet Council of Ministers, signed the Sino-Soviet agreement on economic cooperation between 1959 and 1967 in Moscow. China then sent a group of 14 experts to the USSR to discuss joint aviation projects. In the course of these negotiations, the Soviets proposed the delivery of MiG-21s and also agreed to grant the production licence for the MiG-21 to China. China then began preparations, which quickly came to a standstill after the withdrawal of the Soviet specialists. Then the Soviets made the surprising offer in February 1961 to grant China the licence to produce the MiG-21 regardless of all their differences. Those responsible were undecided.

In view of the urgent needs of the air force, the Vice Chairman of the National Defence Council, Marshal Nie Rongzhen, convened a special meeting of the Central Military Commission on the construction and development of the aviation industry and specifically on the introduction of the MiG-21. The air force and aviation industry reported to Premier Zhou Enlai on the opportunities and risks of the project. It was decided that the Chief of the Air Force, Liu Yalou, should immediately travel to the USSR with a delegation to conduct negotiations on licence production and take the opportunity to familiarise himself with advanced fighter aircraft technology. However, Zhou Enlai made it clear that no contract would be concluded if political conditions were attached to the granting of the MiG-21 production licence. In March 1961, the Chinese delegation, consisting of aircraft, engine, equipment and materials experts, travelled to Moscow.

During the negotiations, the Chinese experts gathered comprehensive information on the MiG-21 and were also given the opportunity to visit the aircraft factory in Gorky (GAZ 21), which was already producing the MiG-21 in series. On 30 March, after several days of negotiations, Liu Yalou signed a contract on behalf of China for licensed production, technical and manufacturing information on the MiG-21 aircraft and the K-13 (R-3S) missile. The agreement stipulated that the USSR would provide China with production drawings and technical documents on the MiG-21F-13, the R11F-300 jet engine and the K-13 infrared-guided air-to-air missile free of charge between August 1961 and October 1962. At the same time, China would purchase raw materials, raw parts and finished products for aircraft, engines and missiles that were not available in China at that time. The USSR agreed also to supply China with a prototype aircraft, two engines and assemblies for aircraft, engines and missiles. The total cost of the deliveries amounted to 90.32 million yuan. At the same time as the material deliveries, 12 complete MiG-21F-13s would be procured for the People's Liberation Army Air Force (PLAAF) to equip the 11th Flight Training School. This unit, founded in 1953, was used to train command personnel and test new models. Its purpose

When the Chinese delegation visited the aircraft factory in Gorky in March 1961, the change in production from the MiG-21F (pictured here) to the MiG-21F-13 had already taken place. Nevertheless, one feature of the MiG-21F, the twin-gun armament, was to become typical for almost all Chinese MiG-21 derivatives. (Holger Müller)

was to familiarise itself with the performance of the aircraft and thus lay the foundations for later deployment with the operational units. After their year of introduction, the MiG-21F-13s were also known as Fighter 62.

For the licence production of the MiG-21, Aircraft Factory No. 112 in Shenyang was selected, which at that time was the most technologically advanced production facility in China. The Soviet design drawings and technological information were delivered there in 1961, totalling 13 tonnes.

According to the Chinese, the scope of delivery did not include design and construction documents for the aircraft, information on fatigue tests, wind tunnel tests and design calculations and this would prove to be a hindrance in later Chinese efforts to further develop the type. However, it can certainly be assumed that the Soviets had provided all the information necessary for the start of licence production, but only this information. The Chinese expectation of receiving the entire development documentation for the aircraft at this point seemed completely unrealistic, both in view of the political situation and in view of the clearly defined task of setting up licensed production.

In view of the difficult situation of the Chinese economy as a result of the Great Leap, a conference of the National Defence Industry Committee at the beginning of 1961 decided to reduce the number of major projects in the aviation industry. Most large and medium-sized projects were suspended and all efforts were concentrated on the licensed production of the MiG-19. As a result, the majority of the employees of the Shenyang Aircraft Factory working on the MiG-21 were also assigned to this project. However, the MiG-21-related technical preparations were not to be completely discontinued. Rather, the aircraft designers of the newly founded Aviation Research Institute were to concentrate on the so-called 'technical penetration' of the MiG-21.

'Technical penetration' was a somewhat more in-depth approach than simple reverse engineering. Through the largely exact reproduction of the object at hand, by analysing the original design ideas, construction methods and technical characteristics, the acquisition of the relevant key technologies and the verification of the findings through calculations and tests, the technology of the MiG-21 could not only be imitated, but actually mastered.

The fact that this effort was being made was not least due to a lesson learnt from the Chinese aviation industry's experience with copies and indigenous developments in previous years. As objective economic and technical laws were disregarded in design and production and exact scientific analysis was replaced by ideology, China had failed miserably in its attempt to achieve the 'most advanced level in the world'.

At the end of 1961, work began on translating, copying and proofreading the drawings and technical documents for the MiG-21 supplied by the USSR. While the aircraft factory in Shenyang – supported by the supplier industry – worked on implementing the production processes on the basis of the Soviet documents, the Aviation Research Institute began to systematically technically penetrate the technology of the MiG-21 from the beginning of 1963.

As a result, problems that had already been encountered during the licence construction of the MiG-19, such as vibrations in the tail section and contamination of the hydraulic system, could be solved. By the end of 1963, the Shenyang Aircraft Design Institute had completed work on the most important aspects.

At the same time, preparations were underway for the trial production of the R11F-300 engine at the Shenyang engine plant, while the Shenyang Aeroengine Research Institute (Institute 606) was responsible for the technical penetration of the engine and was investigating performance development, thermal load, aerodynamic structural analysis and the behaviour of the compressor, among other things.

As a result of all the analyses, it was determined that more than 10,000 components were required for series production of the MiG-21, which could not be produced in China at this time. This also included auxiliary products, weapons systems, etc., required for the manufacturing process, for which appropriate solutions needed to be developed. During more than two years of intensive technical penetration of the MiG-21, a total of 48,000 pages of technical documents provided by the USSR were worked through and information on the design principle of the MiG-21, the key technologies and the necessary raw materials and finished products was collected.

Between April and September 1962, the Shenyang aircraft factory successively received 15 MiG-21F-13s of production lots 16 and 22 in CKD (component knock-down) kits from Gorky, as well as one prototype. When the delivery was inspected it was discovered that rainwater had penetrated many of the crates during transport, due to inadequate packaging, leading to varying degrees of corrosion of numerous assemblies. As a result, anti-corrosion measures were taken and at the same time preparations for assembly were accelerated.

On 23 November 1962, the Chief of Staff of the Chinese armed forces, Luo Ruiqing, approved the assembly of the CKD kits. But it was only after the technical penetration was completed at the end of 1963 that the Shenyang Aircraft Factory officially began assembling complete aircraft from the seven CKD kits of the 16th production batch in February 1964. By the end of March, the first MiG-21F-13 was assembled from the Soviet parts and initially underwent ground testing. As the factory test pilots had no MiG-21 flying experience, pilot Ge Wenrong from the 11th Flight Training School, which was equipped with MiG-21s, was sent to carry out the tests. As the runway at the Shenyang Aircraft Factory was being renovated at the time, he took-off from Anshan airfield on 30 April for the successful maiden flight. By the end of September 1964, the first 10 MiG-21s assembled from Soviet parts had been delivered to the air force. The remaining five aircraft were completed by February 1965.

An elite unit of the PLAAF, the 3rd Air Force Division, was equipped with the MiG-21s assembled from the CKD kits and deployed on combat missions. The unit had relevant experience as it was already equipped with imported MiG-21F-13s in November 1962.

After the Gulf of Tonkin incident in August 1964, the US military began bombing North Vietnam and sent Ryan AQM-34 Firebee reconnaissance drones into China's airspace. These operated at altitudes of 18,000 to 20,000 metres, which was above the peak altitude of the fighter aircraft in the PLAAF inventory. As the MiG-21 could reach a ceiling of 19,800 metres without external loads, the 3rd Air Force Division deployed combat units equipped with MiG-21s to areas where the drones were likely to be used. On 15 December 1964, a MiG-21 fired two K-13 AAMs for the first time at an altitude of 17,700 metres but was unable to destroy the target. After more than a year of unsuccessful attempts to shoot down the Firebees with guided missiles, it was decided to use unguided rockets and cannons and to fire from a relatively close range. The first launch was successful on 3 January 1966, followed by five more by 1968.

During the assembly of the CKD kits, it was realised that the aircraft of the two production batches supplied differed from each other, obviously due to the fact that improvements had been introduced during production in the USSR (a practice that continued throughout the entire production period of the MiG-21). It was therefore decided to use the CKD kit with the construction number 74212210 as the basis for the copy. This resulted in numerous changes to the design. In addition, many new components, materials and technologies used in the MiG-21 were under development due to the still underdeveloped Chinese industry at the time. Numerous changes therefore had to be made to the Soviet design in order to adapt it to China's actual capabilities. In total, the Shenyang Aircraft Design Institute revised the MiG-21 documentation – which comprised almost 50,000 pages – four times. In March 1964, the Shenyang Aircraft Factory began preparations for trial production.

Within six months, almost 20,000 internal process specifications were drawn up and the entire production line was set up. In November 1964, the preparations were so far advanced that production of the first batch of four prototypes could begin. Prototype 0001 was intended for static testing, the other three aircraft for flight testing of the type, which from this point on was designated the J-7 (Fighter-7)

Parts production officially began on 20 November 1964 and, with a total of 22,035 parts, was considerably more complex than for the predecessor J-5 and J-6 models. This was essentially completed in July 1965. The final assembly of prototype number 0001 then took until the end of September. Between 14 October 1965 and 9 January 1966, static strength tests were carried out on the airframe. The entire aircraft was loaded to 112 percent of the design load without any damage occurring. Overall, the test results confirmed the fulfilment of all design requirements. However, the overall weight had increased by seven percent compared to the Soviet MiG-21, which indicated a general oversizing and poorer flight performance.

At the same time as the MiG-21 airframe was being tested at the Shenyang aircraft plant, the Shenyang engine plant (Plant 410) began reverse engineering the R11F-300 engine based upon the 20 sample engines, drawings and technical materials supplied by the USSR and the results of the technical penetration. Tests of the prototype engines began in the spring of 1965 and the first engine assembly was completed in October of the same year. After more than a year of long-term testing and various modifications, the technical requirements were met in December 1966 and the engine, made entirely from domestic materials, was transferred to series production as WoPen-7 (Turbine Jet Engine 7). The first production model rolled off the production line in June 1967, but it was not until a year later that a 50-hour run was successfully completed. It took more than another year before flight testing of the WP-7 and the WP-7A modification for the twin-engine J-8 began. It would be then a whole decade before they were actually ready for service. Nevertheless, China's engine production had made massive progress in manufacturing processes and materials in the appropriation of engine technology. At that time, 166 factories and 23 research institutes from all over the country were involved in the production of the J-7 and WP-7.

The second prototype, 0002, was ready for flight testing on 28 November 1965. On 17 January 1966, the aircraft completed its successful maiden flight with Ge Wenrong at the controls. By 13 April, a total of 29 take-offs and landings had been completed with a flight time of 15 hours and 23 minutes. The aircraft's stability and manoeuvrability were judged to be good and all systems functioned normally. With AAMs, the aircraft reached a maximum speed of Mach 2.02 and a peak altitude of 19,400 metres.

On 28 December 1966, the PLAAF confirmed that the J-7 prototype had met the most important tactical and technical requirements. It therefore recommended that the aircraft be transferred to series production and incorporated into the armament of the PLAAF and PLANAF (People's Liberation Army Navy Air Force). The entire development and testing phase of the J-7 had taken two years and four months and was completed one year and two months ahead of schedule. This was a significant achievement for the Chinese aviation industry, which had just recovered from a chaotic situation.

After completion and testing of the J-7 prototype, the Shenyang Aircraft Factory began low rate initial production (LRIP) of the J-7 at the end of 1967, which ended in 1970. Twenty-three aircraft were built from scratch. All 39 aircraft produced in Shenyang – 23 from the LRIP, 15 assembled from Soviet CKD kits and the prototype

The missing left gun identifies this aircraft as MiG-21F-13. The construction number 1623 usually given for the aircraft indicates a Soviet CKD kit of the 16th Gorky batch completed in China. The brake chute container under the vertical stabiliser was retrofitted. (Chris Lofting)

This aircraft with the serial 86102, which is no longer recognisable here, is said to be the J-7 prototype 0003, one of the first aircraft built from Chinese parts. The restored aircraft, painted white and green, is now on display in a hall of the Beijing Air and Space Museum at Beihang University. (Guido E. Bühlmann)

Twelve MiG-21F-13s assembled from Soviet parts were handed over to Albania as military aid in 1970. Two of these aircraft, parked in a hardened aircraft shelter in Tirana-Rinas, were still airworthy at the time of decommissioning in 2005. (Holger Müller)

aircraft – were handed over to the air force. Twelve of them later went to Albania as military aid.

In the midst of preparations for series production, relations between China and the USSR continued to deteriorate. At the same time, relations with the West also reached a low point during the Cultural Revolution. In this critical situation, the political leadership proclaimed 'survival and security' as the most important priorities. The north-east of the country was seen as particularly vulnerable in a possible conflict with the USSR due to the common border. In order to prevent the destruction of important armaments factories immediately after the outbreak of war, Mao instructed in June 1964 that the most important armaments factories be relocated from the coastal plains to the mountainous hinterland. Even though the strategic benefit of this measure was close to zero, a total of 1,100 industrial plants were relocated. The relocation also affected the production of the J-7 and its further development. The Chengdu aircraft plant (Plant 132), whose construction was essentially complete at this point, was to act as the lead site for the further development of the aircraft and the Guizhou aircraft plant (Plant 162) would serve as a second, quasi-standby production site. The Shenyang aircraft plant would support the establishment of production at both sites with design drawings, process information, sample parts and material samples, and production was transferred step-by-step to the new sites.

In parallel to the transfer of J-7 production to Guizhou, the production of the WP-7 engine was transferred from the Shenyang engine plant to the Guizhou Liyang engine plant (Plant 460), a branch of the Guizhou aircraft plant. Not least thanks to the support of the Shenyang engine plant, trial production could begin in parts of the Guizhou Liyang engine plant at the beginning of 1968, while other parts were still being built. In August 1970, the transfer was completed with a successful endurance test.

However, due to the effects of the Cultural Revolution, the factories in Chengdu and Guizhou were unable to start series production of the J-7 and the WP-7. Immediately after the preparations were completed, production of the J-7 was halted in both factories and the previous production of the J-5 and J-6 continued.

Since the MiG-21F-13 was developed as a high-altitude interceptor, it was designed for high speed and climb performance, which was at the expense of flight performance in the subsonic range and at low altitudes. However, because experience from the air wars of the late 1960s had shown that air combat took place primarily at medium and low altitudes, the air force called for the J-7 to be adapted specifically to the resulting requirements. Aspects of the improvement included safety during ejection, expanding firepower by adding a second cannon and increasing the ammunition supply, improving manoeuvrability and stability, shortening the landing roll distance, increasing engine thrust, reducing fuel consumption and weight and extending the service life.

The fulfilment of such a catalogue of demands ultimately amounted to an extensive redesign and overstrained the still inexperienced Chinese aviation industry, which had been shaken by the Cultural Revolution. Further developments proved this.

In order to fulfil these requirements, the Chengdu Aircraft Factory carried out a series of modifications to the J-7 on a trial basis from April 1969. These included six major changes: The left-hand NR-30 cannon, which was present on the first production version MiG-21F and removed on the F-13, was reinstalled. A thicker intake lip improved the intake behaviour in slow flight. The stepless variable cone adjustment allowed better adaptation of the airflow to the engine's requirements at different flight speeds and altitudes. The modified WP-7B engine had more thrust and a longer service life. Enlarged integral tanks in the wings compensated for the reduction in fuel due to the installation of the second cannon. Finally, the cockpit roof was raised at the rear to create more headroom and – according to Chinese sources – to prevent the pilot's head from hitting it when ejecting. This point is somewhat puzzling, as Chinese pilots were on average smaller than Soviet pilots and the problems with ejecting essentially resulted from the interaction between the canopy and the ejection seat. Obviously, this change was only being implemented on individual aircraft – two of which are verifiable.

To test the planned 'six modifications', the Chengdu Aircraft Factory first modified an aircraft borrowed from the PLAAF. This was said to have actually been a MiG-21, i.e. one of the aircraft delivered directly from the USSR, with the serial 2010. As enlarging the integral tanks in the wings proved to be complex, this was not done and instead the option of carrying additional 480-litre drop-tanks under the wings was created. However, this meant that in this case no weapons (rockets or bombs) could be carried on the wings. Due to technical problems, the modification of the lip of the intake section was also omitted, which meant that the version, the test model of which was completed in April 1969, actually had to be described as 'five modifications'. On 16 June 1969, Wang Ruisheng, deputy head of the 11th Flight Training School, took-off on the first flight with the modified MiG-21 and carried out a total of 28 test flights by September. It turned out that the additional drop-tanks carried under the wings (unlike the wings with two hardpoints introduced later, the drop-tanks were attached to the inner pylon further forward) led to a shift in the centre of gravity, which is why this modification was also discarded. The other improvements were confirmed during the tests, resulting in the J-7I 'four modifications'.

After completion of testing, the Chengdu Aircraft Factory officially began production of five prototypes of the J-7 'six [or four] modifications'. On 12 April 1970, Prototype No. 01 made its maiden flight with test pilot Pang Li-chu at the controls and completed all flight tests by July. Prototype 02 for the static test was completed in August 1970 and subsequently tested at 110 percent of the design load until damage occurred. The J-7 'six modifications' was now officially called the J-7I, while the air force later designated the type as the J-7A. In 1970–71, the Chengdu Aircraft Factory produced a total of 34 aircraft in three batches.

As the factory's technical and technological inspection departments were dissolved during the Cultural Revolution and numerous technical rules and regulations were abolished, chaotic conditions developed in the production organisation and massive problems arose in production quality. Added to this were design deficits and inadequate testing of the aircraft. As a result, the air force refused to accept the aircraft. As the manufacturing plant was unable to make improvements, the 34 J-7Is were parked at Chengdu airfield for an extended period of time.

After analysing the problems that had arisen and in view of the technical expertise available and the production conditions at the manufacturing plant, the original 'six modifications' project was stopped and a simpler version was to be built instead. The modification of the canopy, the installation of the WP-7B engine and the increase in fuel capacity were all cancelled. The installation of a second cannon on the left-hand side of the fuselage, the stepless adjustment of the inlet cone and a thicker and blunter inlet lip were retained.

The prototype of the resulting J-7I 'three modifications' was created by converting one of the J-7I 'six modifications'. Test pilot Jing Zhenghua took-off on its maiden flight in June 1973. In September of the same year, the aircraft was transferred to the China Flight Test Establishment (Institute 630) in Xi'an, Yanliang. A total of 63 test flights lasting 33

The 3487 (construction number 70004) on display in the Datangshan Museum was one of only two known J-7I 'six modifications' (the other with the serial number 9957 is in the National Defence Park in Chengdu) with the raised rear canopy for more headroom. (Guido E. Bühlmann)

hours were carried out there, testing both the general flight behaviour and the cannon armament. At the same time, the Chengdu Aircraft Factory handed over seven examples of the J-7I 'six modifications', which were produced before 1971, to the 7th Air Force Division for operational testing, which successively flew 1,578 missions totalling 925 hours. In addition, the 3rd Flight Test Regiment of the Air Force and the Training Centre (formerly the 11th Flight Training School, and nowadays better known as the FTTC), carried out 417 test flights totalling 193 hours at the Chengdu Aircraft Factory. These tests, which lasted one year, finally completed the trials of the J-7I 'three modifications'. Simultaneous firing of both cannons was possible without interference. The blunt inlet lip reduced losses due to flow separation at low speeds. The stepless variable adjustment of the inlet cone improved the efficiency of the inlet section, increasing engine thrust and acceleration in the supersonic range. The engine, equipment and radio-technical equipment all functioned to full satisfaction.

The small numbers available and the initial problems in production affected the acceptance of the new type by the pilots of the PLAAF, who generally favoured the J-6. To ensure a realistic evaluation of the J-7, the 11th Flight Training School was commissioned in 1975 to carry out a comparative test between the J-6 and J-7. The test revealed advantages for both aircraft in certain aspects, but a general superiority of the J-7 overall. This paved the way for mass production.

On 18 June 1975, the successful completion of the development of the J-7I was confirmed by the Central Military Commission, but at the same time requirements for further improvements were formulated. At this time, the reorganisation of the aviation industry began under the leadership of Deng Xiaoping. This finally enabled the systematic development of the industry on a solid economic and technical basis. However, a number of setbacks had to be overcome first.

In 1975, the Chengdu Aircraft Factory began small series production of the J-7I 'three modifications'. Just under three years later, in February 1978, more than 30,000 defects and problems were identified during quality inspections in production, according to Chinese figures. After a complete reorganisation of the production process, the State Steering Committee for Aerospace Products finally approved the transition from small series production to mass production in May 1980, although some problems remained unresolved.

This step was long overdue, as Deng Xiaoping, then Chief of the General Staff of the People's Liberation Army, had called for the accelerated modernisation of the PLAAF in a speech in October 1977. To this end, the current fleet was to be completely replaced by J-7s and production was to be focussed exclusively on this type. At this point in time, even decisions on individual aircraft types were a matter for the leader in China.

Production of the J-7I continued until 1981. A total of 188 aircraft were produced and 187 delivered, in addition to the 34 of the 'six modifications' variant mentioned above, 153 of the 'three modifications' variant. There is no information on the whereabouts of the missing aircraft, possibly as a test aircraft at the manufacturer. Forty J-7Is were handed over to North Korea as military aid, all others to the PLAAF.

In order to increase the range of the aircraft, an enlarged fuselage drop-tank with a capacity of 720 litres was developed. Five of these were delivered to the air force for testing together with a correspondingly modified fuselage hardpoint. Thus, the original range requirements could be achieved without any design changes to the airframe.

Obviously inspired by the Soviet MiG-21 from the second generation onwards, in which the brake chute container was relocated from the underside of the rear fuselage to the root of the vertical stabiliser, a corresponding modification was implemented on the J-7, albeit with an indigenously developed container with only one door that opened upwards (the Soviet model had two doors that opened sideways). Thanks to this modification, the brake chute could be opened at an altitude of 0.5 to 1 metre before touchdown, which shortened the landing roll distance. On 16 June 1975, the Chengdu Aircraft Factory completed the modified aircraft. In November of the same year, it was trialled in a total of 22 test flights in cooperation with the PLAAF. The modification was judged to be successful and all subsequently produced aircraft were equipped with it. It is likely that the previously delivered aircraft were also retrofitted with the new container, as almost all preserved J-7Is have this feature.

When the design of the J-7I was approved in June 1975, the Central Military Commission realised that the problem of the Soviet original's unreliable escape system had not yet been solved. Parallel to the start of small series production of the J-7I, the Chengdu Aircraft Factory began in July 1975 to eliminate this problem and at the same time improve the tactical and technical performance of the J-7I. The resulting version was initially called the J-7I mod, later renamed the J-7II and designated the J-7B by the PLAAF.

North Korea is the only foreign operator of the J-7I, which was actually only intended for domestic use. As recent pictures show, some of them with the current ruler Kim Jon-Un, the majority of the 40 aircraft originally delivered are still in service. (DPRK/Yonhap)

The most striking external change compared to the J-7I was the completely redesigned canopy. The previous one-piece construction had been redesigned into a fixed front section with a windscreen and two side windows and a rear-hinged, upward-opening canopy similar to that of the MiG-23. In combination with the new Type II ejection seat (HTY-2), an improved escape system was created.

Another significant modification was the first use of the WP-7B engine, here in the batch 02 variant. Development of a more powerful version of the WP-7 for use in the twin-jet J-8 began as early as 1964. The increase in power was to be achieved by using new combustion chamber flame tubes, increasing the turbine inlet temperature and a redesigned afterburner chamber. The higher turbine inlet temperature required cooling of the turbine blades, which were therefore hollow. The resulting technological requirements overwhelmed Chinese industry, which had been shaken by the Cultural Revolution, and was why the development of the new WP-7A version was not completed until 1982. At the same time, the Guizhou Aircraft Engine Factory developed the WP-7B for the J-7, which differed from the WP-7A primarily in the design of the afterburner and the length of the afterburner chamber. After intensive development work and an enormous investment of resources with thousands of hours of test runs on the ground and in the air, a ready-to-use engine was finally available with government certification at the end of 1978, which was then transferred to series production in 1981.

The higher turbine inlet temperature of the WP-7B also increased the exhaust gas and external temperature of the engine, which inevitably lead to heating of the surrounding fuselage structure and parts of the airframe. For this reason, measures were taken to thermally insulate the rear fuselage structure of the aircraft and to actively cool the engine. However, these were obviously not enough, because after several years of operation, the PLAAF discovered that the thermal insulation panels of the rear fuselage of several J-7IIs were burnt and scorched, leading to a temporary grounding of the J-7II fleet. To solve the problem, Chengdu Aerospace Corporation Ltd (formerly Chengdu Aircraft Factory) conducted numerous test flights with aircraft equipped with special measurement technology. Based on the findings, Chengdu modified all J-7IIs in service with the PLAAF and PLANAF between January and May 1992. The solutions found for thermal insulation and cooling were also incorporated into the series production of the J-7 from January 1992.

On 30 December 1978, test pilot Yu Mingwen took-off on the maiden flight of the first prototype of the J-7II. During the test flights from March to June 1979, this and another prototype achieved a flight time of almost 60 hours in 94 flights. Design modifications solved numerous technical detail problems. A static test of the complete airframe, tests of the hydraulic and control system, endurance firing tests of the two cannons and overload landings with two 500kg bombs were carried out successfully. On 26 September 1979, the design was approved by the state authorities. After a small series production of 40 aircraft in five batches, the J-7II was finally transferred to series production in 1982. As more than 80 percent of the individual parts were identical to those of the J-7I, the changes in the production process were limited.

In addition to Chengdu, Plant 162 in Guizhou also started series production. Since 1982, both sites had been producing in parallel with stable quality and increasing quantities. By the end of production in 1986, a total of 375 J-7IIs had been produced, around 100 of them in Guizhou. For the military parade on 1 October 1984 on the occasion

A J-7II of the 44th Fighter Division of the People's Liberation Army Air Force approaching its home base. Clearly recognisable are the slotted landing flaps, which all MiG-21/J-7 versions without SPS boundary layer control system had. (Guido E. Bühlmann)

of the 35th anniversary of the founding of the People's Republic of China, 35 J-7IIs flew in seven groups of five aircraft over Tiananmen Square in Beijing.

In later PLAAF service, the J-7II's fuselage hardpoints were modified to tow the Type-4 aerial target. The aerial target was used for firing training with live ammunition by both fighter aircraft and anti-aircraft artillery.

With the start of Deng Xiaoping's reforms in 1978, the premiss of Chinese politics changed. Ideology receded into the background, while economic aspects became increasingly important, as the desired development of the national economy required considerable investment. The foreign currency required for this was to be procured not least through the export of defence equipment.

This was part of a comprehensive reorganisation of the aviation industry within the framework of the so-called four transformations: from purely military production to mixed civilian and military production, from copy to own designs, from domestic supply exclusively for domestic production to supply for both domestic and foreign production and finally the transition from administrative management (command economy) to management from an economic point of view.

The first customer of the newly founded China National Aero-Technology Import and Export Corporation (CATIC) for defence exports was Egypt, whose air force urgently needed new equipment after the end of the Arab-Israeli war of 1973. In April 1980, Egypt and CATIC agreed on the delivery of 60 F-7s, the first of which were based on the J-7I, modified to meet the requirements of the Egyptian Air Force and designated F-7A. At the end of 1981, beginning of 1982, Egypt received the first delivery of 20 F-7As.

After Egypt received the aircraft and their shortcomings such as low fuel capacity and unreliable escape system became apparent, the recipient country insisted in May 1981 that the remaining 40 aircraft be delivered in an improved version. The resulting modification is largely identical to the J-7II. The Chengdu aircraft factory refers to it internally as the 'J-7II export type', F-7IIC or F-7W, but still classifies it as an F-7A, even though it externally resembles the F-7B. The wiring in the wings was prepared for the use of French R.550 Magic 1 AAMs. Deliveries to Egypt were completed by the end of 1982.

On 27 June 1981, CATIC and the Kingdom of Jordan signed a contract for the purchase of 80 modified J-7II aircraft with an order value of more than US$200 million. The implementation of the improvements requested by the customer proved to be complex and lengthy, which is why the original schedule could not be adhered to. Both parties therefore agreed that an initial 20 aircraft, with only a few modifications would be delivered by the end of 1982. The remaining 60 were to follow after completion of all planned modifications. The first 20 modified J-7IIs were designated F-7Bs.

The Jordanians demanded that Western infrared-guided missiles, an improved escape system and an extended service life of the aircraft equipment. A calculation carried out for the first time by the Chinese industry in this respect resulted in an expected service life of 60 percent of the values stated for the Soviet MiG-21, which was a very low level in view of the known weaknesses of the Soviet equipment. Design and wiring of the wing pylons were modified so that French R.550 Magic missiles or indigenous PL-2s could be used as an alternative. The engine used was the WP-7B (batch M), which had been in series production since the beginning of 1982. The extended service life of the avionics was to be ensured by imported components, although these would only be available for equipping the later F-7M.

The J-7II's ejection system did not fulfil the zero-zero capability demanded by Jordan. After further negotiations, Jordan accepted an alignment to the capabilities of the current Soviet seats, i.e. the ability to eject from zero altitude and in the speed range between 130 and 850km/h. This information is somewhat puzzling since, according to all available photographic evidence, the HTY-2 was always installed in the F-7B, the lower speed limit of which was 250km/h. In order to verify the modifications carried out, tests were performed using an H-5 (Il-28).

The F-7B development programme was officially launched in November 1981. On 16 May 1982, the version successfully took-off on its maiden flight under test pilot Liu Jianzhong. By November 1982, 20 aircraft had been assembled, initially delivered to Jordan and later deployed to Iraq.

In addition to the 40 F-7Bs for Egypt and the 20 for Iraq, Zimbabwe received four and the USA 12 (according to other sources, the USA acquired 15). There is no evidence for the alleged delivery of aircraft of this or a later version to Yemen and Sudan.

3
UNSUCCESSFUL RECOURSE TO SOVIET TECHNOLOGY

Although the J-7II eliminated the main deficits of the J-7 in terms of usability and reliability, the aircraft largely corresponded to the first-generation MiG-21 and thus to the state of the art of the late 1950s.

When the J-7II entered service with PLAAF, McDonnell-Douglas (later Boeing) F-15s, General Dynamics F-16s, Mikoyan i Gurevich MiG-29s and Sukhoi Su-27s were already in service or in the final phase of their development, representing the next-but-one generation of fighter aircraft. As China's incipient reform policy and the opening up of the country made direct contact with Western aviation technology possible, the leadership of the state, the Chinese Communist Party, and the aviation industry realised that its own fighter aircraft were technologically outdated compared to those of foreign countries, especially with regard to aircraft equipment, avionics, and armament.

The first-generation MiG-21 and the J-7 based on it were day fighters that were not suitable for combat missions at night or in bad weather. Unsurprisingly, the air force had demanded an all-weather capability ever since the J-7 was first introduced to service. Following an inspection of the MiG-21PFM equipped with a radar in North Korea, the Chengdu Aircraft Factory began a preliminary study for an all-weather variant of the J-7 in 1972, which was completed in May 1975. Due to the plant's workload with the development and production of the classic J-7, the all-weather J-7 was to be developed in cooperation with the plant in Guizhou. Following an inspection of the MiG-21MF in Bangladesh, Egypt and Tanzania, in 1976 the latter variant was chosen as the basis for the development of the all-weather fighter aircraft, at first designated J-7A.

The initially sluggish development process was accelerated following the Central Military Commission's decision to replace the J-6 with the J-7 and J-8 as quickly as possible, as requested by Deng Xiaoping in October 1977. Detailed planning began in July 1978 and in August and September 1978 two groups of technicians were sent to Romania and Bangladesh, two friendly countries equipped with MiG-21MFs, in order to gain a comprehensive understanding of the technology of that aircraft. At the same time, China succeeded in procuring a MiG-21MF and a MiG-21US from Egypt.

On 3 February 1979, the Egyptian MiG-21MF was transferred to the Chengdu Aircraft Factory. As only a prototype aircraft was available at this time, but no production documentation, the development process differed completely from that of the previous licence production. By precisely measuring and analysing the prototype aircraft, the actual design could be completed relatively quickly. However, it was not possible to analyse the design ideas behind the construction, which limited the knowledge gained with regard to the future development of the aircraft. The dimensional accuracy of this type of reverse engineering was also limited, not least because the prototype aircraft had been repaired after an accident. Between October 1979 and December 1981, 180 employees of the Chengdu Aircraft Design Institute measured the prototype aircraft. Work then began on the prototype of the version now known as the J-7III. Chengdu took over the production of the fuselage, horizontal and vertical stabilisers, nose landing gear and final assembly, and Guizhou the remaining components, including the wings. This was the first time in China that the development of an aircraft had been carried out jointly by two factories. Compared to the original J-7II, the J-7III had significant changes in structure and system equipment. Eighty percent of the parts differed from those of the J-7II.

At the same time as the MiG-21MF was analysed at the Chengdu Aircraft Design Institute, the Guizhou Engine Design Institute and Liyang Engine Plant started preparing to clone the R13-300 engine under the designation WP-13. Measurements and documentation were carried out between July 1978 (probably abroad at first) and March 1979. Subsequently, the Guizhou Liyang Engine Plant and the Chengdu Engine Plant (Plant 420) jointly started trial production. The first batch of three engines was completed in December 1980. As the R13-300 was technically much more advanced than the WP-7 (which in turn was based on the early R11F-300 and did not feature the further developments introduced with the F2, F2S and F2SK sub-versions), the Chinese plants encountered numerous difficulties, with the result that the new engine did not pass the government acceptance tests on the ground until 1985. Another two years later, on 30 November 1987, the development of the WP-13 was finally completed.

In January 1983, parts production for a total of six prototypes of the J-7III began. In October of the same year, assembly of Prototype No. 01 intended for flight testing began. On 17 December, the fuselage manufactured in Chengdu and the wings manufactured in Guizhou were assembled. On 6 February 1984, the aircraft was handed over for flight testing.

In January 1984, Prototype No. 02 was was completed, intended for static test, which it successfully passed on 6 April. As the disassembled prototype could not be restored to flying condition, another MiG-21MF was purchased from Egypt (for US$5.5 million) to compare the J-7III's flight performance with this aircraft. On 19 July 1983, the aircraft, which was given the serial 150, was brought to the Chengdu Aircraft Factory, where it began flight tests on 25 March 1984, piloted by test pilot Yu Mingwen. The autopilot and the blown flaps were tested in another 20 flights, as there was no experience with these components in Chinese aircraft.

On 26 April 1984, Prototype No. 01 took-off on its maiden flight with test pilot Yu Mingwen at the controls. By November 1984, the prototype fleet had grown to five aircraft. Testing was originally scheduled to be completed by December 1985, but defects in various components that had been insufficiently tested on the ground – especially the WP-13 engine – delayed this until October 1987.

Assembly of the first batch of eight J-7IIIs began in February 1988, but the associated WP-13 engines were not initially available. The aircraft could only be delivered to the PLAAF in December 1989, when they were designated J-7C.

Regardless of its ability to intercept targets day and night in all weather conditions, the J-7III's agility and range were reduced compared to the original J-7 due to its modified aerodynamics and higher weight. The increase in the size of the rear fuselage compartment impaired visibility from the cockpit to the rear. The JL-7 monopulse fire control radar could not suppress ground clutter, making it impossible to engage low-flying targets. As a result, the order volume from the air force was considerably reduced and only 17 were produced by 1996. Export versions of the J-7III, which would have been of interest to numerous customer countries in view of the existing MiG-21MF fleets, did not progress beyond an early planning stage.

In view of the gross disproportion between effort and benefit in the development of an all-weather capable J-7, fundamental questions arose: why was a costly parallel development of a completely new type being initiated instead of incorporating findings and solutions

This dismantled Egyptian MiG-21MF was photographed while being transported on a lorry. The prototype aircraft for the J-7III probably arrived in China in a similar condition in 1979. (Holger Müller)

Like Egypt, Bangladesh also had Soviet MiG-21MFs, which was why Chinese experts experts visited the neighbouring country in 1978 to inspect this all-weather version. A few years later, the Soviet MiG-21 versions of the Bangladesh Air Force were replaced by Chinese ones. (Holger Müller)

The J-7IIIA combines the airframe of the MiG-21MF with numerous specifically Chinese systems (clearly visible here are the chaff and flare dispenser under the fuselage tail, the radar warning receiver antennas on the nose and vertical stabiliser as well as the missile rails), but it was still not a success. The appearance of several J-7IIIAs among the exhibits at the museum in Datangshan suggests that the mini-fleet had been decommissioned in the meantime. (dlsunshy)

from the analysis of the MiG-21MF step-by-step into the evolution of the classic J-7? And why were the lower agility and range of the aircraft and the lack of look-down-shoot-down capability cited as reasons for abandoning large-scale production, although these facts must have been known after the inspection of Soviet MiG-21s in friendly countries, or at the latest after the start of test flights with the Egyptian MiG-21MF?

It is possibly the realisation of the considerable technological backlog and the resulting feeling of weakness that led China to pursue two paths in parallel: the further development of the J-7 on the basis of the newer Soviet MiG-21 versions, and the integration of Western technology and its own modifications into the MiG-21F-13-based models. While the first approach ultimately proved to be a dead end in large parts, the second became the starting point for all modern Chinese fighter aircraft development. Since the classic J-7 could be further developed into all-weather fighters and cover both domestic requirements and the export market, the J-7III was ultimately dispensable.

Nevertheless, while working on the J-7IV, aka J-7E, with its numerous innovations, Chengdu Aerospace Corporation was attempting to rectify the problems of the J-7III identified by the air force and improve the aircraft's combat characteristics.

Further development of the J-7III under the direction of Wang Zi-Fang began in 1988, starting with modifications to the J-7III prototype 151. On 20 August 1991, test pilot Liu Qingli completed the first test flight with the aircraft. On 27 November 1994, the aircraft was transferred to small series production as the J-7IIIA and entered series production on 25 December 1995.

Despite the improved performance and maintainability of the J-7IIIA, range and manoeuvrability could not be significantly increased compared to the J-7III, so that only 32 aircraft were produced before production was discontinued in 1999. The version is designated J-7D by the PLAAF. A planned F-7D export version was not realised.

On 1 October 1999, formations of J-7Cs and J-7Ds flew over Tiananmen Square during the military parade marking the 50th anniversary of the founding of the People's Republic of China. Despite the small numbers, which made these aircraft exotic in the fleet, they had been in service for many years.

Regardless of the failure of the J-7III development line, the technology used there was being incorporated into the further development of the classic J-7. The WP-13, developed from the R13 engine of the MiG-21MF, provided a more robust and more powerful engine with which all Chinese single and twin-seat versions, as well as the later JL-9, would eventually be equipped. The twin-barrelled, very compact and therefore easier to integrate GSh-23 cannon of the MiG-21MF would be permanently installed in the two-seaters for export and made available in a podded version for the domestic two-seaters.

4
TWO-SEATER – A PREREQUISITE FOR MASS USE

Like the users of the MiG-21 in the early 1960s, the PLAAF initially had no two-seater aircraft available for pilot training on the J-7. This was not a problem as long as only a few aircraft were in service and being flown by selected pilots with the appropriate flying skills.

However, with the ever-increasing use of the J-7 by PLAAF and PLANAF pilots, training was becoming an urgent problem. Due to the special flight characteristics of delta-winged aircraft, the transition from swept-winged trainers, such as the JJ-5 and JJ-6 to J-7 and J-8, was not only lengthy, but also involved safety risks. In order to shorten the transition and make training more effective, the PLAAF called for an advanced trainer capable of twice the speed of sound based on the J-7. The aircraft must also be able to carry out air and ground combat missions in clear weather conditions during the day and at night. The preparation time for the next take-off should not exceed 30 minutes and the aircraft needed to remain ready to fly for at least three days without maintenance.

In February 1979, a MiG-21US was procured from Egypt to serve as the basis for the development of a J-7 two-seater, at the same time as the MiG-21MF which served as a prototype for the J-7III. Initially, the development of the model designated JJ-7 was prepared at the Chengdu Aircraft Factory, but later handed over to the Guizhou Aircraft Factory. As the MiG-21US was not airworthy, it was dismantled into its main assemblies and shipped by rail from Chengdu to the plant in Anshun, Guizhou province.

The design work for the JJ-7 began in 1981. Probably based on the experience with the copy of the MiG-21MF, where exact measurements and analyses proved to be extremely complex and time-consuming, the Guizhou Aircraft Design Institute focussed on developing an operational prototype as quickly as possible by making the smallest possible design changes compared to the single-seat J-7 and largely adopting existing solutions. The prototype aircraft was therefore not replicated schematically, but the design was merely modelled on it. With the exception of the two cockpits, the airframe structure was essentially the same as that of the J-7II, but the enlarged vertical stabiliser of the MiG-21US was discarded and the tail section of the J-7II was used instead. To improve stability, two large fins replaced the previously central false keel under the fuselage. Also, the aircraft did not receive the SPS boundary layer control system. Equipment that had already been tried and tested continued to be used.

Nevertheless, work progressed slowly as the JJ-7 was not included in the national development plan due to a lack of funds. With 23 million yuan of its own funds, the Guizhou Aircraft Factory prevented development from being interrupted until 1983, when the JJ-7 was finally declared a national key development and development picked up speed. The design was finalised in March 1983. Despite being derived from the J-7II, 40 percent of all parts were changed compared to the J-7II, so extensive aerodynamic, structural and system tests and simulations were carried out.

A total of five prototypes were manufactured at the Yunma Aircraft Manufacturing Plant (Plant 130) and Longyan Aircraft Manufacturing Plant (Plant 150), which belong to the Guizhou Aircraft Factory, as well as by other parts suppliers. Prototype 01, intended for flight testing, was completed on 9 April 1985. On 12 June, Prototype 02 successfully withstood a load of 103 percent of the design load during static tests. After several taxiing tests, Prototype 01 successfully completed its maiden flight on 5 July 1985 under test pilot Yan Xiufu. Prototypes 01, 04 and 05 were transferred to the Xi'an Yanliang Flight Test Institute for state certification, where the flight tests were successfully completed

In addition to the MiG-21MF as a pattern for the J-7III, China also received a MiG-21US from Egypt, which formed the basis for the development of the twin-seater JJ-7. While the exterior of the J-7III was a direct copy of the MiG-21MF, the JJ-7 combined elements of the MiG-21US and the J-7II. (Holger Müller)

on 11 August 1987 after more than two years and 485 flights with a flight time of 286 hours.

Two major problems arose during testing: high cockpit temperatures and longitudinal vibrations in flight. To solve the first problem, the air flow rate of the turbo cooler and the distribution of cooling air in the cockpit were improved. The second was identified as a phenomenon caused by the pilot. No design changes were made; instead, the pilots were instructed to avoid abruptly switching off the afterburner.

On 4 February 1988, the JJ-7 received state certification and the Guizhou Aircraft Factory began small series production. However, more than five years passed before the version was officially transferred to series production in December 1993.

The pilots regarded the JJ-7 as a powerful, reliable and well-maintainable aircraft. However, the performance at low and medium altitudes, the high landing speed and the poor visibility from the rear cockpit were criticised. Solutions to these problems, such as those found on Soviet two-seaters from the MiG-21US onwards with the SPS boundary layer control system and the periscope for the flight instructor, were not introduced on the JJ-7 until the end of production.

Information on the number of JJ-7s produced for domestic use is not available.

CATIC marketed the export version of the JJ-7 under the designation FT (Fighter Trainer) -7. The FT-7 was essentially the same as the JJ-7, with the exception that some equipment was customised and all Chinese lettering was replaced by English lettering. In 1987, the JJ-7 was presented at the 37th Aerosalon in Le Bourget, making it the first Chinese-developed fighter aircraft to be shown abroad. In 1989, the FT-7 flew at the Thai Aviation Festival as the first Chinese aircraft at an international aviation event. On 25 April 1988, the export version of the JJ-7 was certified and on 13 June the first eight aircraft were delivered to Iraq via Jordan under the designation FT-7BI. Subsequently, Pakistan and Iran received four and six FT-7Bs respectively 'on loan', until the versions with extended fuselage were available (see below), Bangladesh three FT-7BBs, Sri Lanka one FT-7BS, Myanmar six FT-7BKs and Zimbabwe two FT-7BZs.

Unlike the JJ-7, at least some FT-7 versions were fitted with a periscope for the rear seat.

In view of the limited operational capabilities of the original JJ/FT-7, Pakistan demanded numerous improvements for its two-seater version based on the single-seater F-7P. In March 1989, the Guizhou Aircraft Design Institute began to realise these under the designation FT-7P.

Externally, the FT-7P differed from the JJ-7 in that it had an extended fuselage, four wing pylons and the twin-barrelled 23mm cannon of the J-7III under the fuselage. The FT-7P had an increased weapons payload, a modernised weapons and fire control system, improved communication and navigation equipment, a radar warning receiver (RWR) and an improved instrument layout in the cockpit. The maintainability of the aircraft was also improved.

As Pakistan was pressing for early delivery, design work and parts production were carried out under high pressure. On 6 August 1990, the Guizhou Aircraft Factory completed final assembly of the first FT-7P and on 9 November 1990 the new version took-off on its maiden flight. Between 1991 and 1993, 20 FT-7Ps were delivered to Pakistan. Certification by the Ministry of Aviation Industry only took place on 13 May 1992, after deliveries had begun.

Iran received a version based on the FT-7P airframe but equipped with much simpler technology, such as the HTY-2 ejection seat, under the designation FT-7N. Twenty-six aircraft were delivered between 1992 and 1996.

With the introduction of improved versions of the J-7 and J-8 into service, aircraft were now in use that had heads-up displays (HUDs) and other new avionics as well as more modern cockpit layouts. The JJ-7 with its electromechanical instruments was therefore only suitable for training pilots for these types to a limited extent. The Guizhou Aircraft Design Institute therefore began modernising the two-seater version in February 1994 under the designation JJ-7A. Among other things, the aircraft was equipped with a HUD, as used in the J-7E. Unlike the export versions, however, the JJ-7A did not have the extended fuselage or the twin-barrelled cannon. However, the latter could be carried in a container on the fuselage hardpoint. The successful maiden flight took place on 24 May 1995 and the design was certified on 7 December 1996. Deliveries to the PLAAF and PLANAF began

The FT-7 was shown at air shows at home (as here in Zhuhai) and abroad. The flight demonstration of an FT-7 at the Thai Aviation Festival in 1989 was the first ever such presentation of a Chinese military aircraft outside China. (Guido E. Bühlmann)

The first two-seaters exported, including the FT-7BS for Sri Lanka, largely corresponded to the JJ-7 for the PLAAF. Modifications to the airframe and equipment were only integrated later at the customer's request. (Erik Sleutelberg)

in 1997. By the end of series production on 26 September 2016, more than 100 JJ-7As had been built. Individual used JJ-7As were exported to Bangladesh and loaned to Pakistan. All other aircraft were flown exclusively by the PLAAF.

Along with the upgraded F-7PG single-seaters, the Pakistan Air Force also received a trainer version with the airframe of the FT-7P and the equipment of the F-7PG, designated the FT-7PG. Unlike the single-seater, the classic delta wing would be retained.

The FT-7PG completed its maiden flight in June 2002. Nine aircraft were delivered to the Pakistan Air Force.

Together with the J-7G-based single-seaters, a number of trainers with corresponding equipment were also exported, which were labelled

Increased fuel capacity and additional wing pylons helped to improve the combat characteristics, and thus also increase the export opportunities, for the stretched FT-7P two-seater. Consequently, the new version was presented at air shows and all subsequent export variants were derived from it. The aircraft in the picture bears the construction number TP0919, which indicates that the construction numbers were assigned consecutively and only the preceding letters distinguish the version. (Guido E. Bühlmann)

FT-7NM, BG, NI, TN and BG-I according to the destination countries. Namibia received two FT-7NMs in 2006, Bangladesh four FT-7BGs also in 2006, Nigeria three FT-7NIs in 2010 and Tanzania two FT-7TNs in 2011. The last of the FT-7P family to be exported are four FT-7BG-I for Bangladesh in 2012–13.

5 BREAKTHROUGH WITH WESTERN TECHNOLOGY

On the basis of the common opposition to the USSR, relations between the PRC and Western countries began to improve towards the end of the period of the Cultural Revolution. With US President Nixon's visit to China in 1972, relations with the United States and other Western countries reached their best level since the founding of the People's Republic. This also brought the procurement of advanced aviation technology from Western countries into the realms of possibility. This, in turn, would then form the basis for China's own developments.

From 2 November to 23 December 1978, a delegation of more than 20 experts led by the Minister for the Aviation Industry Lu Dongchang travelled to Great Britain, France and the Federal Republic of Germany to obtain first-hand information about the aviation industry of the leading Western countries. At the time, both France and the UK were hoping to sell fighter planes to the Chinese, but the prices for these products were too high in view of China's economic situation.

As a result, J-7 and J-8 modernisation groups, formed by the military and industry, decided that a number of avionics and radar components should be imported from the UK in order to eliminate the major weaknesses of the Chinese aircraft and at the same time raise the technical level of the domestic avionics industry through replication and further development.

On 3 March 1979, CATIC and the British company GEC-Marconi entered negotiations on the purchase of avionics components. After 16 months of tough talks, the two parties finally signed a contract on 30 June 1980 for cooperation on the modernisation of the J-7 – the first military-technical cooperation project between China and Western countries after the country's reforms and opening up.

In addition to an agreement for the introduction of six types of equipment from GEC-Marconi, CATIC also signed a contract for the import of radar altimeters from the Smith Company, also based in the UK. The two contracts included 124 sets for equipping 100 J-7s, including the necessary spare parts. At the same time, manufacturing licences are granted for these devices.

The entire avionics suite was designated MADS-7 and consists of the Skyranger 7 radar rangefinder, the Type 956 combined head-up display/weapon computer (HUDWAC) as well as a flight data computer, a multiband radio station, a camera, a radar altimeter and a static inverter (US-made). The latter, which replaced the original rotary inverter, was needed to ensure the necessary quality of power supply for the new equipment.

The imported components were supplemented by new devices from local production, including a compass, radio beacon receiver, pitot tube and angle-of-attack sensor. Modifications to the airframe were also necessary to install the new components. The electrical properties of the inlet cone were adapted to the requirements of the British radar rangefinder. In order to be able to mount the VHF antenna, the upper end of the vertical stabiliser was replaced by a cover made of glass fibre reinforced plastic. To improve maintainability, the pitot tube was shortened, moved from the lower to the upper side of the air intake section and fixed – similar to the Soviet MiG-21 from PF version onwards..

A Chinese pilot flew this aircraft from Dalian airfield to South Korea while testing British avionics. Contrary to what was often reported, this was not a prototype of the J-7IIA, but a J-7I of the early series without the brake chute container at the vertical tail root. (Guido E. Bühlmann)

The significantly more complex equipment of the J-7 required the involvement of new suppliers in production. The Lanzhou Changfeng Machinery Plant (Plant 781) would manufacture the radar rangefinder, the Chengdu Aircraft Equipment Plant (Plant 161) the flight data computer, the Tianjin Radio Plant (Plant 712) the radio station, the Changling Baoji Machinery Plant (Plant 782) the radar altimeter, the Xingping Qinling Aircraft Electrical Plant (Plant 135) the static inverter and the Xi'an Northwest Optical Instrument Factory (Plant 248) the camera. Optical components of the HUD were also manufactured there, while the Suzhou Changfeng Machinery factory (plant 171) was responsible for the electronic components and their assembly.

In September 1980, the Chengdu Aircraft Factory finalised the overall design of the fire control and avionics system of the aircraft known as the J-7IIA. Originally, the plan was to use imported equipment for the first 100 aircraft and then components from domestic production. However, due to a large number of import projects and the resulting strain on foreign currency reserves, the government in Beijing then demanded that import plans in all areas of industry be strictly reviewed and all non-essential projects be cancelled. This coincided with a drastic reduction in military spending. As a result, the PLAAF no longer had the funds to procure foreign components. In February 1981, the decision was taken to discontinue the modernisation of the J-7II.

If the contract was cancelled, the British would have to be paid several million pounds as compensation. At the same time, the opportunity for technology transfer was lost. When Jordan announced the order of J-7s with an improved fire control system almost simultaneously with the cancellation decision, the Ministry for the Aviation Industry decided to implement the contract with Great Britain and to use all the components purchased to equip the export aircraft. The J-7IIA was to be used and further developed as a test aircraft for the project.

Two aircraft from the then-current production of the J-7II were to be used to test the modifications. In addition, J-7I number 139, which was already being used by the factory, was used to test the indigenous components and system modifications. The two J-7IIAs were completed in the third quarter of 1982. After completion of the ground tests, test pilot Yu Mingwen took-off in one of the aircraft for its maiden flight on 7 September 1982. After 15 test flights, the aircraft were transferred to the Xi'an Yanliang Flight Test Institute for further testing in October. As the J-7IIA was the first Chinese aircraft to be equipped with a HUD, four test pilots were sent to the UK for training before flight testing began.

During the test flights, which would later also be carried out at the naval air base in Dalian, numerous problems with the avionics had to be resolved with regard to function and accommodation in the airframe.

On 7 August 1983, a pilot deserted to South Korea in one of the test aircraft. However, this was not one of the J-7IIA prototypes. Instead the pictures from South Korea show an early production J-7I with the armoured glass panel in the cockpit, so this aircraft was probably not equipped with a HUD. And while Chinese sources state that the aircraft was later returned to China, a J-7I with the registration 045 (though not in its original livery) is on display in a South Korean museum today.

The testing of the J-7IIA proved the massive improvement of the aircraft's capabilities in terms of weapons employment, navigation and communications through the use of Western equipment, paving the way for its use in subsequent generations of the J-7. The export version, which combined the innovations of the J-7IIA and the F-7B, was given the designation F-7M. It is the first J-7 version to have four wing pylons and was initially marketed under the designation 'AirGuard'.

Development of the F-7M officially began in September 1981 and thus took place in parallel with that of the J-7IIA and F-7B. During the development process, extensive tests were carried out on the ground and in the air to ensure the reliability required by customers. Two prototypes of the F-7M were completed by August 1983. On 30 August 1983, Yu Mingwen successfully performed the maiden flight. As the British equipment ordered had not yet arrived at this time, the first test flights took place without it. To speed up the flight tests, two F-7Bs were used for test flights with AAMs, to test the cyclic reignition of the engine and measures to prevent an engine shutdown. Two J-7IIAs were used to test the integration of the British equipment. In addition, two J-7IIs were also being used to test the fuel system with additional wing drop-tanks and for weapons firing tests. The eight aircraft carried out a total of almost 500 test flights.

China received massive support from Pakistan for the testing of the F-7M. After the then Deputy Chief of the Air Staff of the Pakistan Air Force visited the test flights in Dalian in July 1983 and expressed interest in the model, Pakistan provided pilots to test and improve the radar with regard to suppressing ground echoes. As a result of the tests, Marconi modified the Skyranger radar rangefinder accordingly.

The F-7M was actively marketed internationally. The cover picture of this CATIC advertising brochure shows the F-7M prototype and the J-7IIA 0139, which was involved in component testing. The latter was later converted into a test bed for the J-7FS and was given a completely new nose section with a fixed nose (see page 28). (CATIC)

Myanmar was the last export customer to receive aircraft with F-7B-generation features. The final F-7IIK/BK is said to have been delivered in 1999. (Alexander Golz)

Due to the imminent delivery date, the Chengdu Aircraft Factory began production of the first 10 aircraft in October 1983, although testing had not yet been completed and the Marconi avionics had not yet been delivered. These did not arrive until March 1984 and in April Chengdu completed final assembly of the first batch of F-7Ms. The first aircraft were delivered in July of the same year. By May 1985, all 60 aircraft had reached Jordan, from where they were transferred to Iraq.

The F-7M family, which also included the later F-7MP, F-7P, F-7MB and F-7N, subsequently achieved notable international sales success with just under 200 units. Financially weaker customers were supplied with hybrid versions that had features of both the J-7II and the F-7M. For example, Zimbabwe received eight F-7IINs in 1987, Myanmar 10 F-7IIKs in 1990 and a further 48 aircraft under the designation F-7BK by 1999, and Sri Lanka four F-7BSs in 1991. According to unconfirmed reports, one F-7M was also delivered to the USAF.

With the F-7M, attempts were made to adjust the weight distribution of the aircraft by using composite materials. A correspondingly manufactured vertical stabiliser section was successfully tested, but not put into series production due to a lack of experience with the durability of such materials.

When the PLAAF had to abandon its plans to modernise its J-7II with the use of foreign avionics for budget reasons in the early 1980s (see J-7IIA), the issue of increasing the combat effectiveness of this version became an urgent problem. A solution was also found abroad in the West.

The overwhelming successes of the Israelis against Syria in the air war over the Lebanese Bekaa Valley in 1982 also impressed the PLAAF. In particular, the modern infrared-guided missiles of the Python series aroused the interest of the PLAAF, which at that time did not have a weapons system with even approximately similar capabilities.

At the same time, Israel hoped to persuade China to reduce its exports of weapons and equipment to Arab states on the basis of improved relations. Based on this constellation, China and Israel began secret negotiations on the Python 3 air-to-air missile in 1982. Although the two countries did not officially establish diplomatic relations, an agreement was subsequently concluded on the licensed production of this weapon, the purchase of a batch of missiles and the necessary production equipment for the China Southern Aviation Power Machinery Co Ltd (Plant 331) in Zhuzhou. This missile has since been manufactured under licence using the designation PL-8.

In August 1983, the Chengdu aircraft factory started development of a J-7II equipped with the PL-8. In March 1985, an accordingly modified aircraft took-off for its maiden flight with Yu Mingwen at the controls. On 9 September 1985, the version designated J-7IIH or J-7HH (with reinforced corrosion protection of the aluminium and magnesium alloys of the airframe for use by the PLANAF) was accepted. In February 1986, units began to be equipped with this type,

Pakistan pushed the development of the F-7 decisively forward through its support for testing, but also through extensive demands for improvements. As a result, the F-7P (pictured here) and the later F-7PG were both versions that were several steps ahead of the then-current domestic models. (Holger Müller)

which was later renamed J-7H by the PLAAF. By the time production was discontinued in 1993, 221 aircraft had been delivered.

After economic constraints had prevented the J-7IIA from utilising Western technology for domestic requirements, high growth rates in the Chinese economy and increasing export surpluses provided the air force with additional funds in the mid-1980s. This made it possible to procure new modern combat aircraft and increase the combat effectiveness of the existing ones. Two corresponding programmes are said to have been implemented from 1986 onwards, although reliable evidence is still lacking. According to various sources, the J-7IIM was created from existing J-7IIs by fitting the combined HUD/956 weapon computer, a top-mounted pitot tube and wings with two pylons each, and with the capability to deploy the PL-7 and PL-8. The J-7IIK, on the other hand, is said to have been built with similar equipment but with a bottom-mounted pitot tube from around 1990 in 59 units.

The information on both projects does not appear conclusive. It is incomprehensible why the earlier modernised version should have had more modern features than the later newly built version. For both versions, there is a lack of photographic evidence confirming the actual deployment.

The improved F-7 also aroused the interest of China's neighbour Pakistan. Following an in-depth inspection of the aircraft during the aforementioned visit by the Deputy Chief of the Air Staff of the PAF in Dalian, Pakistan declared its intention to procure the type as a replacement for the F-6.

In December 1983, the Pakistan Air Force sent two pilots to the flight test centre in Xian Yanliang to test the J-7IIA. After 19 test flights, the Pakistani pilots rated the aircraft's performance as very good but suggested some improvements. Following a request from Pakistan, China transferred two aircraft, a J-7IIA and an F-7M, to Peshawar and Masroor on 9 June 1984. During more than two months of trials, Pakistani pilots took-off on 40 flights and conducted air-to-air gun and missile firing as well as air-to-ground gun, rocket and bomb firing. Simulated dogfights were conducted against F-6s, Mirage 5s and F-16s. The tests were successful in principle and Pakistan subsequently announced the procurement of 55 F-7Ms. However, the sales negotiations between CATIC, the Chengdu Aircraft Factory and Pakistan were protracted due to considerable demands regarding the improvement of the aircraft, so that the first batch of 20 aircraft was not ordered until 1985. Even after that, the delivery status was still the subject of negotiations and was not finalised until November 1987. The version modified in accordance with Pakistani requirements was given the designation F-7MP.

On 9 June 1988, test pilot Lu Guotang took-off on the first flight with the F-7MP. After a handover ceremony on 26 July 1988, the 20 aircraft were transferred from Hotan Airport in the autonomous province of Xinjiang (near the border with Tibet) to Pakistan.

At the end of 1987, Pakistan ordered a further 40 aircraft, but requested improvements to the ejection seat, cockpit layout and equipment. In September 1989, delivery of the upgraded F-7P to the PAF began. In 1993, the PAF decided to replace the previously used radar rangefinder with the Grifo-7 multifunction fire control radar from FIAR (later SELEX Galileo, now Leonardo) from Italy. This made the F-7P suitable for air combat and ground attack in all weathers. An aircraft equipped with the Grifo-7 was tested for the first time in May 1996 and Pakistan procured a further 32 F-7Ps.

In April 1989, China and Bangladesh signed a contract for the purchase of an F-7M derivative, designated F-7MB and modified according to Bangladeshi requirements. The 16 aircraft, which could carry photoreconnaissance pods among other things, were delivered to the Bangladesh Air Force (BAF) via Kunming Airport in October 1989.

Confronted with the MiG-21 in the war against Iraq, the air force of the Islamic Revolutionary Guards Corps of Iran decided to acquire a number of F-7Ms in view of the performance and the low operating costs of this type. The version developed in accordance with Iranian requirements in terms of armament (including compatibility with existing US weapons systems), cockpit equipment and low-cost avionics equipment was designated the F-7N. Between May 1991 and February 1992, Chengdu Aerospace Corporation delivered a total of 30 aircraft in two batches to the Islamic Republic of Iran.

6
DETOURS TO SUCCESS

Driven by both domestic demand and the requirements of its largest export customer Pakistan, the Chengdu Aircraft Factory investigated further approaches to make the J-7 more powerful and, above all, all-weather capable, in parallel with the copy of the MiG-21MF. The biggest problem here was the design of the air intake, which prevented the installation of radar antennas with a larger diameter and at the same time restricted the installation volume for multifunction radars, thus making greater detection ranges impossible.

Following the evaluation of the J-7IIM in 1984, the Pakistan Air Force defined the replacement of the radar rangefinder by a radar with air-to-ground mode as a key requirement for the further development of the type. In addition, the WP-7B engine was to be replaced by the more powerful and efficient American F404, the fuel capacity increased by 1,000 litres, the electronic equipment modernised, the pilot's view improved by a one-piece windscreen and the service life of the airframe extended. The fulfilment of these requirements was equivalent to a completely new design. The J/F-7CP was developed in response to the Pakistani Sabre II tender for the procurement of around 150 aircraft, with 'CP' standing for joint development by China and Pakistan. China's rapprochement with the West in the mid-1980s facilitated cooperation with US companies, and Chengdu and the Grumman Aircraft Company jointly prepared a feasibility study for the Sabre II project based on the F-7CP.

To reduce costs, an aircraft with lateral air intakes and a fixed radar nose was created while retaining as many components of the F-7M as possible. The wing was equipped with a droop nose to increase manoeuvrability at medium and low altitudes. The F404 engine improved the thrust-to-weight ratio and maintainability. The internal fuel supply was increased to 3,000 litres. A twin-barrelled Type 23-3 23mm cannon replaced the previous 30mm cannons, and two additional underwing hardpoints increased the total number to seven (although model photos show only five pylons in each case). The airframe was reinforced to achieve a service life of 3,000 hours, as was the main landing gear, as the take-off weight was increased. The equipment consisted of standard F-16C components.

Development work began in 1985 and by October 1987 had progressed to the point where a full-size mock-up could be produced. As the F-7CP was not considered by Pakistan to be a match for the Indian MiG-29s that had just been delivered and the development costs of around US$300 million were beyond Pakistan's economic means, the Ministry of Defence in Islamabad decided at the end of 1987 to cancel the Sabre II project and buy the F-7P from China instead. The development of the F-7CP was then cancelled. The J-7CP programme, an F-7CP with an indigenous engine and indigenous avionics, also fell victim to this decision.

Nevertheless, Chengdu Aerospace Corporation and Grumman decided to jointly develop a lightweight fighter with improved performance for the world market under the designation Super-7 and based on the J-7CP. Key features included the replacement of the delta wing with a trapezoidal wing with a 40-degree leading-edge sweep and lateral air intakes with a 10-degree outward inclination to ensure good efficiency even at high angles of attack. The rear fuselage and tail unit of the J-7 were retained in order to keep the design effort to a minimum.

The comprehensive sanctions imposed by the West following the suppression of the student protest movement on Tiananmen Square in 1989 forced Grumman to withdraw from the cooperation and the Super-7 and other cooperation projects had to be stopped. Nevertheless, models of the design continued to appear at international air shows years later – possibly in the hope of reviving the project with other partners. This hope was not realised. Nevertheless, Chinese industry had gained useful experience in working with Western partners, which had fuelled the development of both the J-7 and more modern models. However, the further development of the J-7 would again be based on the classic J-7 versions with the fuselage of the MiG-21F-13.

While the development of the J-7I/II and F-7A/B focussed on improving the use and maintenance of the aircraft, the J-7IIA/H and F-7M variants modernised their electronic equipment and weapons systems. The flight characteristics had remained the same at best, and in some cases, such as the PL-8-equipped J-7IIH, they had even degraded.

In order to improve the manoeuvrability of the J-7 at medium and low altitudes and at subsonic and transonic speeds, its range, take-off and landing characteristics as well as its avionics and fire control equipment, thus making it fit for future air-warfare theatres, the Chengdu Aerospace Corporation developed a new type based on the J-7II from 1987 with significant changes, which was initially designated the J-7IV and later the J-7E.

The most significant innovation in the J-7E was the replacement of the original delta wing with a double-delta wing with an enlarged surface area and manoeuvring flaps. The double-delta wing was being developed by Northwestern Polytechnical University in Xian, which specialised in aerospace matters. The typical leading-edge sweep of the MiG-21 of 57 degrees remained unchanged in the inner part of the new wings. From the seventh rib outwards, the leading-edge sweep was then reduced to 42 degrees (OKB MiG had already tested a similar configuration with an outer sweep reduced to 48 degrees, but without increasing the wingspan and without additional flaps, on the Je-6/2 prototype, but did not put it into series production). At the same time, the trailing edge in the outer area was given a negative sweep. For the first time in a Chinese aircraft, the leading edge of the outer wing section was fitted with a powered droop nose and the landing flaps were no longer designed as slotted flaps. In addition, the original ailerons were moved outwards by one rib spacing, so that the wingspan and wing aspect ratio were increased. The wing area was increased and thus the wing loading was reduced. The modifications improved both the lift coefficient of the wing and the slow flight characteristics of the aircraft. They also eliminated the problem of the centre of gravity shifting when carrying the PL-8, so that the counterweight in the nose used since the J-7IIH was no longer necessary.

Newly developed aluminium-lithium alloys with higher strength than conventional duralumin were used for the airframe of the J-7E. The extensive use of these alloys for load-bearing structures and surfaces reduced the weight of the aircraft by 17 percent.

The engine used was the WP-13F. The J-7E was the first variant of the MiG-21 or J-7 ever to be equipped with a pressure refuelling system. However, the J-7E still did not have a full radar, but only a radar rangefinder, despite corresponding requests from the PLAAF. The reasons for this were the expected impairment of manoeuvrability due to the installation of a correspondingly heavy radar and the associated considerable modifications to the air intake to accommodate a larger

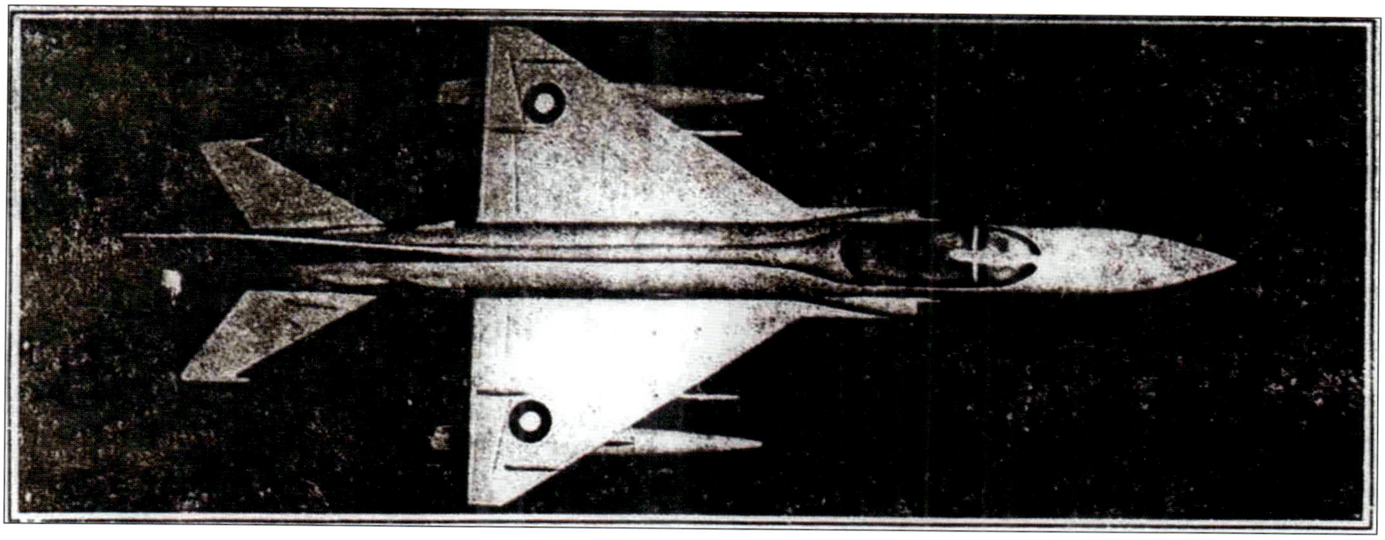

The cover page of the document produced by Grumman for the Sabre II project shows both the high degree of adoption of design features from the J-7 – wings, centre fuselage and tail unit – and the target customer, the Pakistan Air Force, whose nationality markings can be seen on the wings. (Chinese Internet)

Even after the end of cooperation with the USA following the suppression of the student protest movement at Tiananmen Square in 1989, the Chinese aviation industry attempted to market the Super-7 concept developed jointly with Grumman. However, all corresponding activities, such as here at the MAKS in Zhukovsky in 1993, remained unsuccessful. (Holger Müller)

From this perspective, significant differences between the second (F-7BS, right) and fourth F-7 generation (F-7GS, left) can be seen. In addition to the shape of the wings, these include bottom-mounted or top-mounted pitot tube, different shapes and positions of the antennas and retrofitted (on the outside of the rear fuselage) or integrated (in the false keel) chaff and flare dispensers. (Holger Müller)

Thanks to its improved low-speed characteristics, the J-7EB was the first J-7 version to be used by the 1st August Aerobatic Team of the PLAAF. Between 1995 and 1999, the unit flew in the red and white livery shown here. (Holger Müller)

In 1999–2000, the J-7EBs were replaced by new aircraft, or maybe simply repainted. The blue and white livery was retained even after the change to the J-7GB in 2004 and was only changed to blue, red and white when the J-10 took over as a display aircraft. (Guido E. Bühlmann)

antenna. The aircraft would also not be equipped with medium-range AAMs.

The J-7E's manoeuvrability at medium and low altitudes, take-off and landing behaviour and range had been significantly improved compared to the J-7II. Together with improved avionics and weapons payload, the J-7E represented significant progress, but was still far from the state-of-the-art.

Development of the J-7E officially began in October 1987. In March 1989, the design work was completed and the production of four prototypes began: one for static testing and three for flight testing. On 11 March 1990, prototype 01 was completed for flight testing and completed its maiden flight on 18 May 1990 under test pilot Qian Xuelin. The WP-13F engine intended for the J-7E completed its ground testing in April 1992, with a successful test flight in May, and was transferred to series production in September 1992. After three years of flight testing, delivery of the J-7E for troop trials began in August 1993 and series production started in December. More than 260 J-7Es and J-7EHs (naval version) were built by the time it supposedly ended in 2001, which seems inconclusive given that production of the successor version did not begin until 2006. This satisfied the most urgent needs of the air force and older types, such as the J-6 and J-7I, could be decommissioned.

With the availability of the more modern avionics of the J-10, a number of J-7Es were modernised under the designation J-7L.

The PLAAF '1st August' aerobatic team was also equipped with the J-7E. Founded in 1962, the formation initially flew with Soviet MiG-15bis, later on J-5s, J-6s and, from 1980, JJ-5s. When the JJ-5 was due to be replaced in 1992, the K-8 Karakorum trainer and the J-7E were shortlisted. The latter promised considerably higher flight performance, but the team's pilots doubted its suitability in view of the original J-7's poor manoeuvrability at medium and low altitudes.

To dispel these doubts, the Chengdu Aircraft Factory provided the team with first one and later a second J-7E from May 1992. Test pilot Lei Qiang demonstrated the aircraft. The team's pilots then undertook 30 test flights. As a result, the J-7E was finally selected as the successor to the JJ-5. Chengdu developed a corresponding demonstration version, the J-7EB, which made its maiden flight on 2 December 1994.

In the course of 1995, the team took delivery of 12 demonstration aircraft and on 7 July of that year it made its first public appearance at Yangcun airfield in Tianjin with the new design, painted red and white in the tradition of the formation. A colour change to blue and white took place in 1999–2000. The 12 aircraft were said to have been replaced by new ones, but there is no evidence of this in the form of construction numbers and it is possible that the aircraft were only overhauled and repainted. At the end of their service (at the latest when they were replaced by the J-7GB in 2004), they were converted back into combat aircraft.

With the availability of the F-7MG, China launched a new marketing offensive to boost fighter aircraft exports, which had slumped in the meantime. Here the prototype 0144 is seen at its first public demonstration in Zhuhai in 1996. (Guido E. Bühlmann)

At the beginning of the 1990s, after the major orders from Pakistan had been processed, Chinese fighter aircraft exports came to a virtual standstill. This prompted the Chengdu Aircraft Factory to step up its sales efforts. After the end of the 1991 Gulf War, a new round of the global arms race began and many developing countries had a need for more modern combat aircraft but were unable to procure Western models for economic and/or political reasons. These countries were potential customers for a new version of the J-7, which – now that the Super-7 had fallen victim to the political climate – must be based on a different platform.

Thanks to its good performance at low and medium altitudes combined with high air-to-ground firepower, the J-7E was the ideal aircraft for this purpose. In May 1993, after the development of the J-7E was completed, the Chengdu Aircraft Factory defined the necessary technical improvements for an export version of the J-7E. This version was given the designation F-7MG, with the 'G' indicating 'improvement'. The main difference between the export version and the J-7E was the integration of Western avionics – mainly supplied by GEC-Marconi (now BAE Systems) and AlliedSignal (now Honeywell).

In order to market the F-7MG internationally, a prototype was built and made available to potential buyers for testing. The development risks and costs were shared by Chengdu Aerospace Corporation, CATIC and Guizhou Liyang Engines, who signed an agreement in 1996. The contract partners invested 30 million yuan in this project. Eventually, three prototypes – 0142, 0143 and 0144 – were used to test the design. While 0143, originally the prototype of the J-7M, retained the original delta wing and was used to test radar and avionics equipment, 0142 and 0144 were given the double-delta wings of the J-7E and were used for avionics tests as well as for flight tests and demonstrations.

At the first Airshow China in November 1996, the still unfinished prototype 0142 was displayed in one of the halls and the fully modified 0144 was demonstrated in flight. The presentation aroused the interest of the air forces of Zimbabwe, Pakistan, Bangladesh and Egypt, among others, who subsequently trialled the F-7MG prototype. In July 1997, the PAF pilots alone took-off on 12 flights in Chengdu to investigate the flight behaviour of the new version in comparison with the F-7P. However, the F-7MG would not be built in series. Instead, it would form the basis for future export models. And once again, it was Pakistan that provided the decisive impetus for its development.

In the 1990s, there was an ongoing conflict between Pakistan and the USA over Pakistan's nuclear weapons programme. As this made new purchases of US defence equipment unlikely, but at the same time the PAF was still in need of modernisation, China and Pakistan officially began negotiations in 1997 on the F-7PG project, a version tailored to Pakistani requirements that combined the technology of the F-7P with the airframe of the F-7MG.

The negotiations took a long time as Pakistan changed its catalogue of requirements several times. Chengdu carried out more than 300 test flights to select the radar alone. In addition, Pakistani test pilots visited China several times to carry out test flights. It was not until August 2000 that the final choice of radar was made and Pakistan again opted for the Italian FIAR Grifo-7, which was already being used in the F-7P. However, at least the first aircraft would be delivered without the fire control radar and would only be retrofitted with it in Pakistan. The most striking external feature of the F-7PG was the new one-piece windscreen, which not only provided the pilot with improved visibility and situational awareness, but also simplified the use of the HUD. It is unclear why the problem of poor forward visibility, which pilots had complained about since the first MiG-21 versions, was solved so late for the J-7 family (and of all MiG-21s, only the final modernisation variant MiG-21-93 for India – also known as the MiG-21bis UPG – had this improvement). The F-7PG was powered by the WP-13FI engine, which has been in series production since 1995.

In July 1999, even before the final contract was signed, Chengdu began preparing for production. One year later, final assembly began and the successful maiden flight of the F-7PG took place on 20 December 2000. When the PAF signed the delivery contract on 23 June 2001, the first 20 aircraft had already been completed and were

In the course of their service with the PAF, the F-7PGs not only received new liveries several times, including various squadron insignia (here of the 23 Air Superiority Squadron, stationed in Quetta), but also new equipment, recognisable by modified antennas. (Awais Lali)

The only J/F-7 single-seat version with a fixed radar nose that actually flew was the J-7FS. Initially equipped with a delta wing and three-part strutted windshield, later it was fitted with a one-piece windscreen, the double-delta wings and a striking modification to the upper edge of the vertical stabiliser. Here the aircraft takes off with two PL-8 practice missiles. (Chinese Internet)

delivered at the end of 2001. The Pakistan Air Force purchased a total of 48 F-7PGs.

Parallel to the development of the F-7MG/PG, Chengdu again attempted to equip the J-7 with a fixed radar nose. The fact that only a few years after the cancellation of the J-7CP/Super-7 such an approach was pursued again suggests that the Chinese industry was not yet able to realise the desired radar range in the available installation volume. Under the designation J-7F, a J-7 with a fixed radar nose and an air intake under the fuselage, similar to the American F-8 Crusader, was investigated in 1995. The J-7FS, which unlike its predecessor reached prototype status, was developed in a similar design as an industry development initiative. The J-7II prototype 0139, initially equipped with the original delta wings and later converted to double-delta wings, was fitted with a modified nose section and underwent extensive testing until 2003. The result was the F-7MF. With its angular air intake under the fuselage, it bore similarities to the J-10, which was under development at the time. Like the J-7F, the F-7MF did not go beyond a study, as the FC-1/JF-17, a completely new type with a fixed radar nose, was already under development at the time. Although this was the second failed attempt to bring a new nose design for the J-7 into series production, there would be a third, successful one – in the form of the two-seater JJ-7B, also known as the JL-9 or FTC-2000.

As a result, the final J-7 version for domestic use and export also had the distinctive feature of the entire MiG-21 family – the central air intake with a movable cone. The J-7G was based on the J-7E but featured more than 30 improvements that made it the most potent J-7 variant of the PLAAF. Instead of the previous radar rangefinder, a small-sized multifunction pulse-Doppler radar had finally been installed, which gave the aircraft a real all-weather capability and did not impair manoeuvrability thanks to its low additional weight. Together with a helmet display and the one-piece windscreen also used here, the PL-8's ability to engage off-axis targets could be fully utilised. The J-7IIH, which could also deploy this missile, was limited to a significantly smaller operational range of the missile. The WP-13FI engine was also used for this version.

Assembly of the prototype began in March 2002 and the successful maiden flight took place on 28 June 2002. In the same year, two prototypes were delivered to the air force for unit trials. Development was completed in July 2004 and the first series aircraft was delivered in November. Production continued until around 2010. A total of up to 130 aircraft were said to have been produced, although it is not clear whether this figure included the export versions.

Based on the J-7G, the J-7GB demonstration aircraft was created, and replaced the J-7EB in the 1st August team in 2004. The first public flight demonstration with the new version – in front of foreign military attachés – took place on 15 June 2005 at the home base of the Yangcun unit in Tianjin. The J-7GBs would remain in service until 2009, when they were replaced by J-10s.

There is confusion surrounding the F-7 export versions NM, BG, NI, GS and TN, which were being delivered to Namibia, Bangladesh, Nigeria, Sri Lanka and Tanzania respectively. This concerned both the initial basis – F-7MG and J-7G – and the specific equipment.

The F-7GS for Sri Lanka represents the single-seater generation based on the F-7MG or J-7G with a real fire control radar. Obviously the landing flaps, which are extended to 25 degrees in slow flight, are no longer designed as slotted flaps. (Alexander Golz)

With the delivery of 12 F-7BG-Is to Bangladesh in May 2013, production of the J/F-7 single-seaters in Chengdu came to an end. The 2722 with the construction number 1163 was the last aircraft built there. (Holger Müller)

The single-seat J-7E/G versions for domestic use and F-7G for export are the only MiG-21 variants that have a pressure refuelling system. This reduces the refuelling time from approximately 15 to six minutes. (Holger Müller)

Ultimately – as is also known from Soviet production – innovations were constantly being transferred to production. Furthermore, customer requirements as well as the economic and operational possibilities of the customer determined the equipment. Six F-7NMs were delivered to Namibia in 2005 and 12 F-7BGs to Bangladesh in 2006. Nigeria signed a purchase contract for 12 aircraft in 2005, but financing problems delayed delivery of the F-7NI until early 2010. Sri Lanka ordered four F-7GS in 2007 and Tanzania 12 F-7TNs in 2009. Deliveries took place in 2008 and 2011, respectively.

The final and best-equipped export version of the J-7 is the F-7BG-I – also the last MiG-21 single-seater ever built worldwide. Although various sources claimed that the 'I' stands for improved, the aircraft are referred to as 'BG-one' by the BAF. Compared to the F-7BG, they have more modern avionics, including larger colour displays in the cockpit and a helmet display as standard. The contract for the procurement of 16 aircraft was concluded in 2011. Production of the single-seater aircraft in Chengdu ended with delivery in May 2013.

Although numerous new models were now coming off the production line in China's aircraft factories in significant numbers, the procurement of the J-7 for the PLAAF would continue for some time. In September 2016, the completion of the last JJ-7A for domestic use was finally announced. After 57 years this also marked the end of MiG-21 production worldwide. It speaks in favour of the quality of the original Soviet design that the airframes of the last aircraft built largely resembled those of the first delivered.

7
WITH NEW AERODYNAMICS INTO THE TWENTY-FIRST CENTURY

When the most modern versions of the J-7 and J-8 were equipped with HUDs, multifunctional displays, fire control radar, etc. in the late 1990s, pilot training changed from pure flying training to training in the use of these systems. At the same time, the influx of modern models such as the J-11 (Su-27) and J-10 began, these differed considerably from the J-7 in terms of aerodynamics, manoeuvrability, avionics and cockpit equipment. The number of available two-seaters of these types was small and the flight-hour costs were high. The PLAAF was therefore looking for a universal third-generation advanced trainer on which a large part of the flying training could be completed and which would allow quick transition to the operational type with just a few flying hours on the two-seater. It should also be both cost-effective and have modern equipment to enable training in the use of modern avionics for all types of the fleet.

The JJ-7, with its largely outdated equipment and the only slightly improved JJ-7A were not suitable for this purpose. Therefore, in January 2001, the Guizhou Aircraft Factory and the Guizhou Aircraft Design Institute began preliminary studies on their own initiative for an advanced two-seater based on the JJ-7. Initially designated FTC (Fighter Trainer China)-2000, the aircraft was later given the name JL-9 'Mountain Eagle'. The design phase started in December 2001 and the new aircraft was already strikingly different from the other JJ-7 versions, as well as from all other MiG-21s and J-7s, being the first production model of this family to feature side air intakes and a fixed nose. Thus, it was not until the last modification of the MiG-21 series that modifications were introduced which the designers of the original model had already considered a few years after the first flight in the form of the Je-8, later pursued again with the MiG-21Sh, which were never transferred to series production, but which – as became clear during the Soviet comparison flights with the US F-5 in the mid-1970s – would certainly have improved the performance of the MiG-21. As with the J-7E/G, a double-delta wing was used, which differs from that of the single-seat versions, however, in that the outer section is less swept and the trailing edge is straight.

Between September 2002 and October 2003, a prototype (421) was built by Yunma Aircraft Manufacturing Plant and Shenyang Aircraft Manufacturing Plant, which took-off for its maiden flight on 13 December 2003 with test pilots Qian Zhaoquan and Ma Xinxinin in Anshun. This and another prototype (422) had completed a good 300 flight hours in 412 flights by November 2005. Two more prototypes (423 and 424) were added later. Certification took place on 22 December 2005 and small series production for unit trials began in 2007. Series production did not start until 2013. The aircraft was designated JJ-7B by the PLAAF.

For pilot training for carrier operations, the further developed JL-9G was created, which was first mentioned in the Chinese media in 2011, from which the FTC-2000G export version presented in 2018 was derived. While the original JL-9 and FTC-2000 versions have the vertical stabiliser of the JJ-7, the JL-9G and FTC-2000G features a redesigned tail unit with a straight leading edge and larger wing area. In return, the two stabilising fins under the rear of the fuselage have been omitted. The wings of both versions have been completely redesigned. This means that only the rear fuselage, tailplane, the arrangement of the landing flaps and ailerons, the landing gear and the engine are based on the original model. Nevertheless, it is remarkable that aircraft with MiG-21 genes are still being built more than 60 years after the start of production. The only export customer for the original FTC-2000 was Sudan with six aircraft. Myanmar received six, or even 12, of the subsequent FTC-2000G.

Demonstration of one of the FTC-2000 prototypes at the Airshow China in Zhuhai. The similarities (rear fuselage and vertical stabiliser) as well as the differences to the JJ/FT-7 (fuselage nose and side air intakes) are clearly visible from this perspective. (AVIC)

Published on the occasion of the 100th anniversary of the founding of the Chinese Communist Party, this propaganda photo clearly shows the differences between the wings of the J-7G (right) and the JL-9 (left). The two leading edge sections of the wing of the J-7G have sweeps of 57 and 42 degrees respectively, and the trailing edge has a negative sweep of about 9 degrees in the outer area. On the JL-9, the inner wing is also swept at 57 degrees, but the outer wing is only swept at 37 degrees, which, in conjunction with a straight trailing edge, results in a significantly deeper outer wing and a larger wing area, which is additionally enlarged by the leading edge root extension. (Chinese Internet)

8
SUCCESSOR OR NOT

The J-7 was replaced as a Chinese export fighter aircraft by the JF-17 Thunder, which was developed in cooperation with Pakistan. After the Super-7 programme ended for political reasons, those responsible in the aviation industry realised that retaining the airframe of the J-7 would limit the performance of a future type too much and that Pakistani requirements in particular could not be met. At the same time, the collapse of the USSR once again opened up the possibility of drawing on the expertise of OKB MiG. With its support, a type designed exclusively for export was created from the early 1990s under the designation FC-1 or JF-17. Nevertheless, it took until 2003 for the prototype of this aircraft to make its maiden flight.

To date, however, the JF-17 has not been able to match the export success of the J-7. So far, apart from the main customer Pakistan, only Nigeria and Myanmar have procured a few aircraft. The fact that China opted for the J-10, a much more powerful model, for its domestic requirements, while not using the JF-17, did certainly not help to attract potential customers. In addition, the price of the aircraft, which is quoted at US$17–25 million, depending on the equipment, prompted some potential customers to buy cheaper models such as the FTC-2000G (for which the price is said to be only US$8.5 million) or the L-15. It is therefore quite likely that the F-7 – like the original MiG-21 – will remain the most successful export fighter aircraft of its country of manufacture.

This FTC-2000G prototype presented at the Airshow China 2018 has all the characteristic features of this version: negatively swept, diverterless air intakes, wings with a curved inner section and a straight, 37-degree swept outer section as well as a straight leading edge of the vertical stabilizer and missing stabilizing fins under the fuselage. (Open source)

The conceptual successor to the J/F-7 as a light export combat aircraft is the JF-17 Thunder, developed and produced jointly with Pakistan. As this model, on display at the Paris-Le Bourget aerosalon in 1995 shows, the design originally known as the FC-1 was created at a time when the further development of the J/F-7 family was still in full swing. (Holger Müller)

The JF-17 is a completely new design that had no J/F-7 genes and is considerably more powerful but still cannot match the export success of the J/F-7. Instead, JF-17 features such as the 'bumper' air intakes are used in the further development of the FTC-2000 into the FTC-2000G. (Alexander Golz)

9 VARIANTS

The J/F-7 can be divided into three generations based on the MiG-21F-13 airframe, a separate development line based on the MiG-21MF airframe and a series of two-seaters. The latest versions within this development series differ considerably from the original design.

The first generation consists exclusively of day fighters, without fire-control radar and with radar rangefinder, one or two 30mm cannons as well as bombs, unguided rockets and guided missiles on two wing pylons.

The second-generation day fighter features more modern avionics, a revised escape system including a new canopy as well as improved operational behaviour and better maintainability. Later export models of this generation can carry bombs, unguided rockets and guided missiles on four wing hardpoints. Some of them are equipped with a fire control radar and thus have all-weather capability.

The third generation comprises the J-7III/IIIA all-weather fighters based on the MiG-21MF, which are equipped with a fire control radar, a 23mm cannon and armament consisting of bombs, unguided rockets and guided missiles on four wing pylons.

Finally, the fourth generation consists of day and all-weather fighters with modernised avionics and double-delta wings, cannon armament as well as bombs, unguided rockets and guided missiles on four wing pylons.

For the basic and advanced training of pilots, several variants of two-seater trainer versions with and without fire control radar were also produced, some of which are equipped with cannon armament (a 23mm cannon, permanently installed or in a pod on the fuselage hardpoint) and which can carry bombs, unguided and guided rockets on two or four wing pylons. All two-seaters have the original delta wing.

Even if the Chinese MiG-21 derivatives did have their own independent technical solutions, the technologies used were fundamentally based on foreign developments that were copied in China with or without a licence. The engines and equipment in particular can largely be traced back to Soviet and Western systems.

First Generation: Type 62 (MiG-21F-13)

Aircraft assembled from Soviet CKD kits are referred to as Type 62. In fact, they are MiG-21F-13s from Gorky production.

The aircraft is 13.46 metres long without pitot tube (15.76 metres long with pitot tube), 4.10 metres high and had a wingspan of 7.15 metres. The fuselage has a diameter of 1.24 metres and the diameter of the air intake section was 0.69 metres. The empty weight is 4,871kg, the normal take-off weight 7,110kg and the maximum take-off weight 8,625kg. The MiG-21F-13 reaches a maximum speed of M 2.05 (2,150km/h) at an altitude of 12,500 metres and can climb at a maximum rate of 130m/s. The static ceiling is 19,000 metres and the range with an additional drop-tank was 1,450km.

The wings, with an area of 23 square metres, have a triangular (delta) shape with a sweep of 57 degrees, slotted landing flaps with an area of 0.94 square metres each and a deflection angle of 24.5 degrees as well as two wing pylons. The vertical stabiliser has an area of 4.08 square metres.

The inlet cone is set to three different positions depending on the speed – below Mach 1.5 and above Mach 1.9 as well as between these values. The pitot tube is mounted under the air intake and can be swivelled upwards on the ground. The canopy opens forwards and works together with the SK (*Sistema Katapultirowania*) ejection seat during ejection. The brake chute is located in a container on the left under the fuselage. Two narrow airbrakes are integrated into the cannon fairings on each side of the fuselage.

The MiG-21F-13 is powered by an R11F-300 engine. It has a fuel capacity of 2,480 litres, divided into 10 tanks.

The nose landing gear is equipped with a 500 x 180mm KT-38 wheel. The uncontrolled nose wheel deflects solely through the one-sided deceleration of the main landing gear wheels when braking with the pedals actuated. The KT-82 main landing gear wheels measure 660 x 200mm.

The MiG-21F-13 is armed with a fixed 30mm NR-30 cannon in a fairing on the right under the cockpit. The two wing pylons can carry R-3S infrared-guided missiles on APU-13 launchers, bombs weighing up to 1,000kg, S-5 unguided rockets in UB-16 containers or S-21 rockets individually on PU-21 launchers. An additional drop-tank can be attached to the fuselage hardpoint if required.

J-7

The J-7 is a largely unmodified copy of the MiG-21F-13, but features a number of minor modifications, mainly resulting from the production process. Due to technological and design deficits, the aircraft is also a good 300kg heavier than the Soviet original. The avionics components also do not achieve the already modest performance of the original model.

Twelve J-7s (or Soviet MiG-21F-13 assembled in China) were delivered to Albania. There is no evidence that these aircraft received the export designation F-7, as often stated. In Albania, the aircraft were referred to as MiG-21 by the personnel, in the technical documents also as Type 7. The latter and the serial numbers of the aircraft, which are probably also their construction numbers, suggest that they are J-7s from Chinese production.

J-7I (J-7A)

With the J-7I, Chinese designers took the first steps to adapt the J-7I to the needs of their own air force. Two subvariants of the J-7I were created, 'six modifications' and 'three modifications', the first of which contained the modifications requested by the PLAAF and the second the modifications that could actually be realised by Chinese industry.

For the 'six modifications' variant, a Type 30-1 30mm cannon including an ammunition supply of 60 rounds is installed on the left side of the fuselage, which, however, reduces the internal fuel quantity by 100 litres. The air intake cone is now adjusted continuously instead of in three positions and the leading-edge radius of the intake lip is increased from 0.5mm to 2.0mm. A WP-7 engine is built in. The capacity of the integral tanks in the wings is increased to compensate for the loss of capacity due to the installation of the second cannon. The rear part of the cockpit roof is raised by 70mm, giving it a greater angle of inclination. The ejection seat used is an unchanged copy of the Soviet SK.

In the J-7I 'three modifications', only the second cannon, the stepless variable adjustment of the inlet cone and the redesigned inlet lip remain from the above-mentioned modifications. In the course of production, a 720-litre drop-tank and a brake chute container on the vertical tail root were added, which all subsequent versions also received. This allows the landing roll distance of the J-7I to be shortened to 800 metres. The PLAAF referred to the J-7I as the J-7A.

Externally the J-7 does not differ from the MiG-21F-13. The fact that the Albanian aircraft were kept operational through the installation of Soviet-built engines from the mid-1990s onwards underlines that the original type and the licensed version are largely identical. (Holger Müller)

Clearly visible from this perspective is the most striking external distinguishing feature of the J-7I from its predecessor: the second 30mm cannon on the left side. (Marcus Fülber)

F-7A

The F-7A is the export version of the J-7I with the same airframe, including canopy and ejection seat, and with the WP-7 engine, whose service life is increased to 200 hours between overhauls. Some of the electronic equipment has been modified to meet customer requirements. The CL-2 radar rangefinder is replaced by the more powerful Type 222 and the Type 602 friend-or-foe identification device (IFF) by the YD-3, specially developed for export. All Chinese inscriptions on the aircraft are replaced by English ones.

Second Generation

J-7II (J-7B)

The outward distinguishing feature of the J-7II compared to the previous versions is the rear-opening, significantly lighter cockpit roof, part of the most important innovation of this version, the modified escape system. Its second main component is the Type II or HTY-2 ejection seat based on the Soviet SK. Like the old roof, the new one is opened and closed using two air cylinders. However, these are mounted at the rear of the cockpit. Unlike the previous solution, the new combination of seat and roof ensured a safe ejection from the aircraft in an emergency at speeds of 250–850km/h and between 0 metres and ceiling height. The design differs both from the Soviet KM-1 seat with a side-opening canopy and from Western ones of the time, which China did not have access to. The J-7II is equipped with a WP-7B 02 batch engine. The other changes compared to the J-7I are less relevant to performance, but mainly concern technological aspects of a now better-controlled production.

F-7B

The export version of the J-7II is designated the F-7B and, like the original model, has the HTY-2 ejection seat and the rear-opening canopy with fixed windscreen. Unlike the F-7A, the 720-litre auxiliary fuel tank can be carried on the fuselage hardpoint. According to various sources, the early examples of this version which were delivered to Egypt and the USA are designated as F-7IIC (C stands

The F-7B was the first aircraft to fulfil customer requirements to a limited extent, for example with regard to the capabilities of the ejection system. It was not until years later that versions were produced that were fully equipped according to the recipients' specifications. The Egyptian F-7Bs – here inside a hardened aircraft shelter at Tanta – were the first aircraft of this version to be delivered. (Lon Nordeen)

The F-7IIK/BK for Myanmar is a hybrid version of the J-7II (recognisable by the bottom-mounted pitot tube) and the F-7M (wings with two pylons each). The equipment consists of Chinese components. (Dr István Toperczer)

for export) or F-7W (W stands for foreign trade) and sometimes also called F-7A.

The most important equipment components of the actual F-7B are the Type 222 radar rangefinder, the 602 or YD-3 IFF and a Hangjia-10-11 gun-camera as well as a static inverter of US origin. With the new WP-7B engine (batch M), the operating time between overhauls is increased to 600 hours.

In addition to the indigenous PL-2 AAMs, at least the Egyptian aircraft are prepared for use with the French R.550 Magic AAM. Since, according to Egyptian operational experience, the R.550 generates significantly more exhaust gas than the PL-2, which can cause the engines to flame out, the engines are equipped with a device for continuous reignition. A change from the previous petrol starter system to a kerosene starter eliminates the need to carry and handle a separate fuel.

The last F-7Bs exported, such as those for Zimbabwe, Sri Lanka and Myanmar, combine features of the J-7II and the subsequent F-7M. Externally visible are wings with two pylons each, to the outer of which either drop-tanks or armament components can be attached. The fuel system thus also corresponds to that of the F-7M. The engine used is the WP-7B (M batch), while the F-7IIN/F-7BN for Zimbabwe uses the WP-7B (BM batch) of the F-7M. To reduce costs, the American inverter is replaced by a Chinese DBL-500A. Due to the tropical climate in Zimbabwe, the F-7IIN has a cockpit air conditioning system with increased cooling capacity, while the alcohol de-icing system for the windscreen, the heaters for the battery hatch and helmet and the IFF system were omitted. On the F-7IIK/F-7BK for Myanmar, Chinese avionics from the J-7IIH partially replaces those of the F-7M. Aircraft built after 1991, such as the F-7BK, already had the improved heat protection in the rear section, while those built before that were retrofitted – as was previously the case with the J-7II.

The F-7BS for Sri Lanka is essentially the same as the F-7BK but had a WP-7B engine (BM batch). For cost reasons, parts of the equipment, in particular the HUD, were omitted, which is why a Chinese windscreen could be used instead of an imported one.

J-7IIH/J-7H/J-7HH

The upgraded J-7IIH is equipped with new GJ3-D universal wing pylons and a modified control unit for launching the PL-8, alias Python 3, as well as the PL-2 and PL-5.

At 120kg, the weight of the PL-8 is significantly greater than that of the PL-2, which weighed only 75kg, and the associated rearward shift of the centre of gravity has an impact on flight stability. As changes to the arrangement of the fuel tanks were not sufficient to compensate for the shift in the centre of gravity, the trim weight in the aircraft's nose had to be increased to 200kg. The associated increase in mass impairs the manoeuvrability of the aircraft and thus compromises the advantages of the PL-8. In order to be able to accommodate the higher weight, the structure of the wings and the main landing gear strut are reinforced and the main landing gear wheel is replaced by a tubeless LS-16E type. The J-7IIH again has the WP-7B engine (BM batch) with kerosene starter and extended service intervals.

Type 941-4C chaff and flare dispensers are installed in the lower tail section of the J-7IIH, making this J-7 version the first with self-protection devices. The J-7IIHH (later J-7HH) variant of the naval aircraft has the same equipment as the J-7IIH of the PLAAF but is better protected against the effects of sea salt thanks to corrosion protection measures and the replacement of aluminium alloys with magnesium alloys. J-7IIs that are still in service today for training purposes have been fitted with new ejection seats.

J-7IIA/F-7M

Since the J-7IIA, originally developed for domestic use, did not go into series production, but at the same time formed the basis of the F-7M, both versions are discussed together here.

The J-7IIA and F-7M are based on the J-7II airframe but are equipped with Western avionics and new armament. At the heart of the avionics equipment are the GEC-Marconi Skyranger 7M radar rangefinder and the Type 956 HUDWAC from the same manufacturer.

Other new components of British origin include a Type 50-048-02 digital air data computer, a 2032 gun-camera linked to the HUD and an AD-3400 multiband radio station with encryption. The DH-1030-24-1200-CS-IIB static inverter from Phoenix Aerospace and a radar altimeter with a range of 1.5km are manufactured in the USA. The overall avionics system is designated MADS-7. As the quality of the Chinese windscreens is not sufficient for the use of the HUD, this component comes from the British manufacturer Pilkington. The newly developed CW-1002 air data sensor and the Type 602 digital IFF device are manufactured domestically. According to Chinese sources, the service life of the indigenous avionics components was said to be 75 percent of the original Soviet ones. If this is true, 30 years after the development of the original components, which were not particularly durable anyway, this is evidence of a still low level of quality.

External distinguishing features of the F-7M compared to the predecessor models are the broad antenna behind the cockpit and, above all, the pitot tube mounted on the top of the nose and the air data sensor on the left side of the nose, as on the Soviet original.

The F-7M is the first J-7 series version to have two additional outer wing hardpoints (the variants based on the J-7II with four pylons and the J-7III would not be built until later). These can carry 480-litre auxiliary tanks, which together with the fuselage drop-tank significantly increase the range of the aircraft. Alternatively, bombs or unguided rockets can be attached to the outer pylons. As with the F-7B, the inner wing pylons are designed and wired to optionally accommodate PL-2 and PL-7/R.550 Magic 1 AAMs. Alternatively, bombs and pods for unguided rockets can also be attached there.

Due to the higher loads on the wing, its structure is revised, as are the nose and main landing gear. The nose landing gear used is that of the J-7III, under development at the same time. The legs of the main landing gear are reinforced and tubeless tyres used. Due to weight reductions in the airframe, the lighter avionics and the additional wing hardpoints, the centre of gravity of the aircraft shifted to the rear. This is compensated for by a ballast weight of 130kg in the nose.

The engine used is the weight-reduced WP-7B (BM batch), which had been in series production since the end of 1982. The ejection seat is the same HTY-2 as in the F-7B.

J-7IIM/J-7IIK

With the J-7IIM and J-7IIK versions, some of the features of the F-7M find their way into the J-7s of the PLAAF. The Type 956 HUD extends the range of armament applications. The inner wing hardpoint, a GJ3-D universal pylon, enables the carriage of PL-2, PL-5B, PL-7 and PL-8 AAMs, while the J-7II's YD-3 IFF system is retained for domestic operations. The original WP-7B engine is used and the HTY-2 ejection seat also remains unchanged from the J-7II.

F-7MP

The F-7MP, initially designated Skybolt and destined for Pakistan, features numerous modifications in terms of armament and equipment compared to the original F-7M. The F-7MP is equipped with various US-made devices, such as the AN/ARC-164 and AN/ARC-186 radio stations and the AN/APX-101 IFF detection system, as well as the Chinese-made LJ-2 omnidirectional RWR. The cockpit equipment is adapted to the modified avionics. Instruments, switches and levers on the instrument panel and the side consoles are grouped together according to the type and task of the equipment for more intuitive operation. The geometry of the control stick is adapted to the stature of Pakistani pilots. The installation of mirrors in the cockpit improves visibility to the rear. The oxygen supply system is compatible with the Western pilot helmets used by the Pakistan Air Force. The HTY-2 ejection seat with which the aircraft were delivered is replaced in Pakistan by a British Martin-Baker Mk.10L zero-zero seat.

The primary sensor of the F-7MP was a Skyranger 7M radar rangefinder, which is replaced by the Grifo radar of the F-7P in PAF service. The converted aircraft are also designated F-7P. The F-7MP is retrofitted with a Chinese helmet-mounted sight.

The armament of the F-7MP features numerous improvements over the F-7M. The inner wing pylon can carry PL-5B, PL-7/R.550 Magic 1 and AIM-9P Sidewinder AAMs, LAP-68 rocket pods and Mk.82, Mk.20 and Mk.117 bombs. With the exception of the PL-7/R.550 missiles and the Mk.117 bombs, the same loads can be attached to the outer pylon, alternatively so can 480-litre drop-tanks. An additional 480- or 720-litre drop-tank can be carried on the fuselage hardpoint. Maximum take-off weight and maximum speed close to the ground increase. The software of the HUD is adapted to the extended range of weapons.

F-7P

Unlike the F-7MP, the F-7P initially had the Chinese HTY-4 ejection seat, later replaced by the Martin-Baker Mk.10L. The HUD is extended and the instrument scales are changed from metric to imperial units. The antennas of the RWR are repositioned (at the top of the vertical stabiliser) and its previously separate displays were integrated into the HUD to improve situational awareness. A new

Delivery of the improved F-7P to Pakistan began with serial 90521. However, this version would only reach its full performance a few years after its introduction with the installation of the Grifo-7 radar. (Holger Müller)

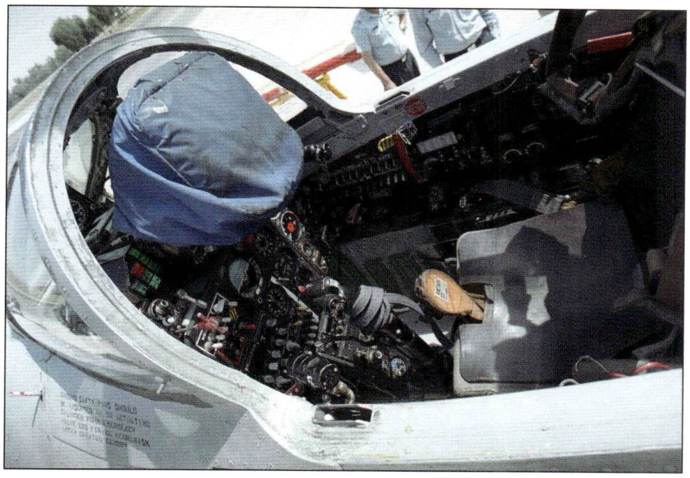

The hosts did not want to show the author a key feature of the F-7P, the HUDWAC, during his visit to the PAF base at Mianwali. However, it is easy to see that the Chinese designers had retained the pin in front of the cockpit that signals the landing gear extension. (Holger Müller)

turbo radiator and additional air outlets enhance the effectiveness of the cockpit air conditioning. The cockpit layout is changed again compared to the F-7MP and later adapted to this variant in the F-7MP. The avionics equipment is said to now include an AN/ARN-147 VHF omnidirectional range beacon and instrument landing system (VOR/ILS), an AN/ARN-149 radio compass (LF/ADF) and a Pro Line II Collins Aerospace distance measuring equipment (DME). However, it is possible that these devices were already present on the F-7MP.

To facilitate adjustment to the pilot's height, the rudder pedals are equipped with a quick-adjustment mechanism. All other equipment and systems remain essentially unchanged compared to the F-7MP.

The F-7P is delivered with versions of the WP-7B engine as well as with the WP-7C.

F-7MB

The F-7MB is essentially the same as the F-7M but has the WP-7C engine. The inner wing pylons are wired for use with the PL-7 air-to-air missile. It is possible to mount rocket pods on the inner and outer pylons. Some aircraft can carry photoreconnaissance pods under the wings. As an alternative to the drop tank, the fuselage hardpoint can accommodate the towed 3A target for aerial gunnery training. The LJ-2 omnidirectional RWR is installed to improve the self-protection capability. There are two variants of the antenna arrangement – on the lateral thickenings of the vertical stabiliser (similar to the Soviet MiG-21) or on its upper edge. Originally equipped with the HTY-2 ejection seat, the aircraft later received the Martin-Baker Mk.10L as part of a modernisation programme.

F-7N

In the F-7N, the radio compass is replaced by the AD2780 tactical air navigation (TACAN) system with air-to-ground function, which improves the accuracy of the aircraft's short-range navigation. The cockpit instrument displays are converted from metric to imperial units. To reduce costs, the old HTY-2 ejection seat and the F-7M's oxygen system are used instead of the F-7P's more powerful oxygen system.

The avionics equipment consists entirely of Chinese components. The WP-7C engine is identical to that of the F-7MB. Weapons computers and pylons are compatible with the US weapons available to the Islamic Republic of Iran Air Force (IRIAF), such as unguided 2.75-inch rockets, Mk. 82 and Mk. 83 free-fall bombs and Mk. 82 Snake Eye retarded bombs. The aircraft are prepared for the deployment of the French R.550 Magic AAM and its Chinese counterpart, the PL-7C.

The F-7MB 1427 shown here carries the antennas of the LJ-2 RWR at the top of the vertical stabiliser. On other aircraft, these are located on the thickened section above the roundel. (Holger Müller)

Although only one wing pylon is fitted here, the F-7N, like all variants of the F-7M, have two hardpoints per wing as standard. The picture was taken on the occasion of the annual Army Day parade in April, when the participating aircraft took-off from Tehran-Mehrabad. (Babak Taghvaee)

Third Generation
J-7III (J-7C)

The J-7III is an outwardly almost identical copy of the MiG-21MF. The equipment consists of copies of Soviet components as well as indigenous developments based on Eastern and Western systems.

The most striking external difference to the previous J/F-7 versions is the enlarged air inlet section, the diameter of which has increased from 663mm on the J-7II to 870mm. The JL-7 single-pulse radar operating in the 2cm band is installed in the inlet cone. In combination with the HK-03D optical sight, the JL-7 can be used day and night as well as in complex weather conditions. In order to ensure sufficient course stability with the thicker fuselage, the fin area is increased from 4.45 square metres to 5.34 square metres. The thickened fuselage spine behind the cockpit accommodates the newly added rear tank, but at the same time restricts the view to the rear. A periscope is mounted on the cockpit roof to enable the pilot to observe the rear airspace. The manually operated canopy opens to the right.

The cockpit is equipped with red light illumination for night flight operations. The escape system consists of an HTY-4 ejection seat, which ensures a safe escape at altitudes from zero metres to ceiling height and at speeds of 130 to 1,000km/h. The larger main landing gear wheels to withstand the increased take-off weight required the enlargement of the wheel wells in the fuselage, the corresponding bulges in the outer skin and the wheel doors.

The engine used is the WP-13, a copy of the R13-300 of the MiG-21MF. The twin-barrelled 23mm Type 23-3 rapid-fire cannon with 200 rounds of ammunition is installed under the fuselage. A 480- or

Twelve J-7s were donated to Albania by the People's Republic of China. Armed with Chinese-made PL-2 infrared-homing, short-range air-to-air missiles (ASCC/NATO-reporting name 'AA-2 Atoll'), these were the only missile-armed fighters in Albanian service after previously delivered Soviet MiG-19PMs were exchanged for Chinese aircraft. Due to chronic lack of funding, they were flown very little: over their 20-years of service most of the airframes clocked between just 300 and 600 hours before being withdrawn from use. Albanian J-7s were painted in light blue-grey overall, while national insignia was applied in six positions. (Artwork by Tom Cooper)

Following 16 F-7MBs acquired in 1989, 12 F-7BGs purchased by Bangladesh in 2006 were originally painted in two shades of blue, as shown on the cover of this volume. During their overhauls and upgrades, the entire fleet was repainted in the livery shown here, consisting of two shades of blue-grey. Older F-7BGs remain relatively easy to recognise from their subsequent brethren, not only due to different antennas, but also because they have the left 30mm gun. (Artwork by Tom Cooper)

JJ-7A 2437 is the last short fuselage trainer procured by the Bangladesh Air Force and was delivered second-hand in 2007. It is shown in its latest configuration, following an overhaul, repainting and installation of additional navigation equipment. (Artwork by Tom Cooper)

Originally, both F-7As and F-7Bs were painted as shown here, in light grey overall, with a disruptive camouflage pattern in dark grey applied on top front, centre and rear surfaces only. This F-7B – serial number 4529 – is equipped with the Chinese-made 720-litre drop tank under the centreline and armed with US-made AIM-9J or AIM-9N Sidewinder air-to-air missiles. No picture evidence is available for the F-7B being equipped with Sidewinder, this artwork shows the modifications that would have been required for such an installation. (Artwork by Tom Cooper)

This F-7A of the Egyptian Air Force (serial number 4506) is shown in the livery applied following overhauls and worn during the last decade of the type's service in that country: with front and rear fuselage and the fin painted in dark ghost grey, and with the entire spine, most of the fin, and the outside half of the wings in dayglo orange outlined in black. In addition to two internally-installed 30mm cannons, primary armament consisted either of French-made Matra R.550 Magic 1 infrared-homing, short-range air-to-air missiles, or the Chinese copy PL-7. The last surviving F-7s were withdrawn from service in Egypt between 2015 and 2017. (Artwork by Tom Cooper)

All Iraqi F-7Ms wore the same camouflage pattern shown here, consisting of beige, dark earth and dark green on upper surfaces and sides, and the Chinese variant of light admiralty grey on undersurfaces. National markings were applied in six positions, and fin flashes in their usual place. The serial number of this example – 1598 – is indicative of it being one of those that reached Iraq at the time, the air force was still applying serial numbers in sequence of their delivery (dating back to 1931). (Artwork by Tom Cooper)

In 1989, the old system was replaced by a new, five-digit system, where the first two digits denoted the type of the aircraft. In this case, this was 21059 (where '21' stood for 'MiG-21' and '509' was the aircraft's individual number). This F-7B was captured by US troops during the invasion of 2003, while still equipped with a Chinese-made 720-litre drop tank. Although armed with two internally-installed 30mm guns, and compatible with a wide range of conventional ordnance, in Iraq the type never saw any combat during the war with Iran in 1980–1988, nor any subsequent conflict. (Artwork by Tom Cooper)

The large number of F-7B/Ms and FT-7BIs, as seen here, enabled the Iraqi AF to vastly expand its fighter weapons school. Designated the Flight Leaders School, this included at least six operational conversion units (Nos. 6, 7, 27, 57, 117, and 131), some of which flew F-7s in addition to older MiG-21 variants, BAC/BAe Jet Provosts and Hawker Hunters, and were supervised by up to 20 instructors detached from the Indian Air Force. F-7s remained in Iraqi service until at least 1997. This jet – serial number 21079 – was one of two eventually captured intact by US troops in 2003 (the other was 21071). (Artwork by Tom Cooper)

Impressed by the intensive deployment of MiG-21s by the Iraqi Air Force, when aiming to establish its own flying sub-branch, the Islamic Revolutionary Guards Corps (IRGC) placed a large order for F-7Ns. The plan did not work out due to the lack of funding and qualified personnel, and they were delivered to the 'regular' Islamic Republic of Iran Air Force (IRIAF), instead. Originally wearing the livery depicted here – including blue-grey and dark sand – the first 14 F-7Ns were assigned to Tactical Fighter Base 5, Omidiyeh, in south-western Iran. This was denoted by the '5', applied in light blue, inside a circle on the fin, below the fin flash. (Artwork by Tom Cooper)

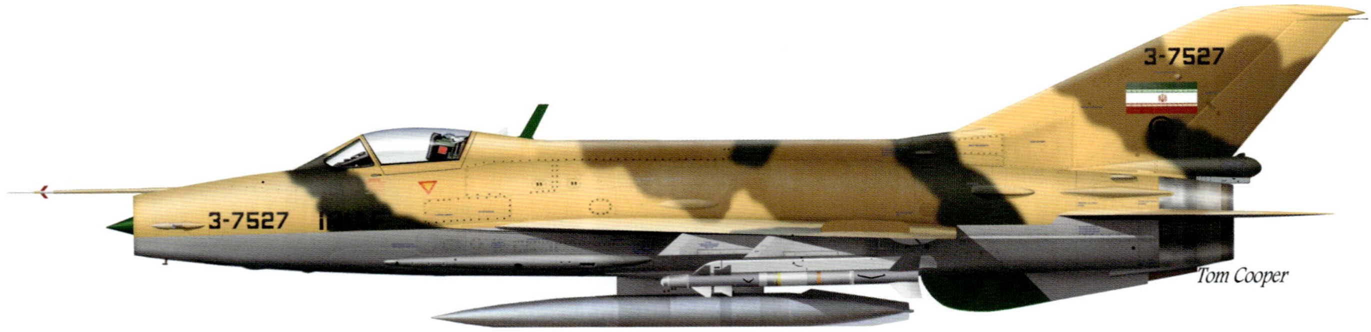

Retaining its standardisation to US-norms, when 30 F-7Ns were overhauled in Iran, they were not only repainted in the same colours as used on the F-5s (tan, dark brown, dark green on upper surfaces and light grey on undersurfaces), but these were also applied following the same pattern. In addition to their internally-installed 30mm guns, the primary armament of Iranian F-7Ms are Chinese PL-7 air-to-air missiles. (Artwork by Tom Cooper)

While initially operating FT-7Bs, the IRIAF experienced a large number of problems related to their poor manufacturing quality. Therefore, the IRIAF was happy to acquire the much improved FT-7Ns as soon as that variant became available in 1992. Assigned to TFB.5, they wore the same livery as the single-seaters. (Artwork by Tom Cooper)

Once all the initial teething problems with FT-7Ns were sorted out, the type was highly popular and became the principal advanced jet trainer and conversion trainer of the IRIAF. However, by that time the IRGC had assumed control of TFB.5, forcing the IRIAF to redeploy its entire F-7N/FT-7N fleet to TFB.8 outside Esfahan. There, the fleet was reorganised into the 84th Combat Training Squadron. Meanwhile, Iran obtained the rights to undertake complete overhauls locally and thus as the jets were subjected to these, they received a new F-5E/F-like livery. Since the early 2010s, this is frequently replaced by very different variants of a 'digital camouflage pattern' one of which is shown on this example (often enhanced with a large 'dragon' insignia, applied in red, not shown here). (Artwork by Tom Cooper)

Myanmar received its first 10 F-7IIK single-seat interceptors (serial numbers 1601–1610) and two FT-7BK two-seat conversion trainers (serial numbers 1611 and 1612) from the PRC in 1991. They were followed by a batch of 48 F-7BKs (serial numbers 1613–1637 and 1639–1641) and four FT-7BKs (1638, shown below, and 1663–1665) in 1999. They all wore a camouflage pattern consisting of dark brown and dark green on upper surfaces and sides, and light admiralty grey on undersides. Original armament included PL-2As: ultimately, these were replaced by the much more advanced, all-aspect PL-5s, illustrated here and – reportedly – various weapons of Israeli origin. (Artwork by Tom Cooper)

All six FT-7BKs eventually delivered to the *Tamdaw Lay* were rather varied in their appearance and equipment. Although applied with many variations, the camouflage pattern essentially remained the same as on the single-seaters: worn-out portions of the lower fuselage and underwing hardpoints were frequently repainted in a stronger version of light blue. The type served as an advanced trainer for two generations of Myanmar's pilots, but attrition was heavy. (Artwork by Tom Cooper)

Myanmar was the launching customer for the latest FTC-2000G: six examples were delivered in late 2022 – all wearing this striking camouflage pattern, very similar to that applied on MiG-29s and Su-30s of the same air force. The type has seen intensive combat service in the ongoing civil war in Myanmar and two are known to have been shot down or lost in flat spins: the example here crashed in January 2024 during a clash with insurgents of the Kachin Independence Army. According to so far unconfirmed reports, Myanmar received six more of these aircraft in 2024. (Artworks by Tom Cooper)

As clear from the number of antenna around their lower forward fuselage and the spine, F-7NMs made for Namibia were delivered to a very high standard of equipment. Rather unusual is their camouflage pattern: consisting of beige and a rather strong shade of light green on upper surfaces and sides, and light blue on undersides, this is highly polished – in turn demonstrating a dramatic improvement in their manufacturing quality in comparison to examples exported 20 years earlier. National markings are applied on the rear fuselage and the fin only. F-7NM serial number 0315 was written off in a landing accident on 15 October 2021. (Artwork by Tom Cooper)

Delivered from 2005, Namibian F-7NMs arrived together with a sizeable assortment of modern Chinese-made weapons. This included (from left to right) PL-9C air-to-air missiles, 250kg bombs, and HL-7 pods for unguided 80mm rockets. To date, there is no photo evidence that the Namibian F-7NM is equipped with the 720-liter drop tank. All relevant photos show the aircraft without or with two 490-liter tanks on the wings. While shown with only one underwing pylon, the F-7NM can be equipped with four under each wing. Their service introduction developed very slowly, and two Chinese pilots flew a pair over the capital, Windhoek, during a military parade on 23 June 2006. (Artwork by Tom Cooper)

To help work up its No. 23 Squadron, Namibia also acquired two FT-7NMs (serial numbers 0330 and 0331). The sole Namibian F-7 unit eventually became fully operational in 2009–2010, by when – thanks to the acquisition of Hongdu K-8 jet trainers in 2006–2007 – enough native pilots were qualified to man all 12 aircraft. (Artwork by Tom Cooper)

Starting in mid-2010, the fighter-jet fleet of the Nigerian Air Force was completely rebuilt with 12 F-7NIs and three FT-7NIs. Their camouflage scheme consists of dark olive green and dark green-brown on upper surfaces and sides, and light blue on undersides. National markings were applied in six positions. As usual on exported F-7s, maintenance stencils – applied in red (warnings) and blue or black (maintenance instructions) – were all in English. (Artwork by Tom Cooper)

The 12 Nigerian F-7NIs were delivered with equipment to a similar standard as those for Namibia. Their armament was similar and including (from left to right) HL-7 rocket pods, Chinese-made 250kg bombs, South African-made copies of the US-designed Mk.82, and also PL-9C infrared-homing, short-range air-to-air missiles. Attrition was heavy: two F-7NIs were written off by May 2011, one crashed in 2015, two collided and crashed in 2018, and the surviving aircraft were all in storage by 2018. Following reactivation and overhauls, another crashed in 2023, prompting Lagos to search for a successor. (Artwork by Tom Cooper)

Originally, Nigeria acquired three FT-7NIs for the conversion training of pilots for the 101st Air Defence Group at Makurdi Air Base. All three two-seaters from the original delivery, including NAF812, have since been lost. Another two-seater (serial number NAF 815) was donated by China free of charge in 2022. Although the NAF has meanwhile started acquiring CAC JF-17 Thunder multi-role fighters, surviving F-7NIs and FT-7NIs remain the backbone of the force, and both versions have seen heavy utilisation in operations against the Boko Haram terror organisation. (Artwork by Tom Cooper)

Export customers for the F-7 family of aircraft. (Map by Anderson Subtil)

A comparison of the cockpit of J-7III (left) and MiG-21MF (right) reveals similarities, but above all significant differences. Particularly noticeable is the omission of the radar display screen on the J-7III. The target information is displayed in the HUD and thus much more ergonomically than on the Soviet original. (Holger Müller collection/Holger Müller)

720-litre auxiliary tank can be mounted on the fuselage hardpoint. PL-2 and PL-5B AAMs, HF-16A or HF-7B rocket pods or, alternatively, 10 100kg bombs, four 250kg bombs or two 500kg bombs can be attached to the four wing hardpoints.

The equipment includes the KJ-11 autopilot, the 506 VHF radio, the 264 A radar altimeter, XS-6A radio beacon receiver, WL-7 radio compass, 930-2 stern RWR, 605 IFF device and Type 481 data transmission/guidance system.

J-7IIIA (J-7D)

For the J-7IIIA, the equipment of the J-7III is further developed and new technology available at the time of development is integrated. The new WP-13FI engine improves the weight-thrust ratio and flight performance.

The HK-13A HUD replaces the HK-03D optical sight and, in combination with the JL-7A fire control radar, forms an integrated fire control system. Other new or modified components include the KJ-11A autopilot, the ADS-1 air data computer, JD-3II TACAN, the 563B inertial navigation system, the WL-7A radio compass, the 265 radar altimeter, the TKR-122 ultra-short wave radio, the self-protection system consisting of the 930-4 omnidirectional RWR and 941-4A chaff and flare dispenser and the Type 23-3A twin-barrelled 23mm cannon. The J-7IIIA can carry the PL-7 and PL-8 infrared AAMs.

Fourth Generation

J-7IV (J-7E/J-7EH/J-7L)

The J-7E again features the fuselage design of the classic J-7 with the narrower air intake section. At the same time, the new double-delta wing is introduced as a significant aerodynamic improvement. The inner wing has a leading-edge sweep of 57 degrees, while that of the outer wing is reduced to 42 degrees and the trailing edge of the wing has a negative angle of 9.37 degrees. This increases the wingspan to 8.32 metres and the wing aspect ratio from 2.2 to 2.8. To improve the low-speed flight characteristics, the wings have droop noses with a maximum deflection angle of 25 degrees on the outer wing and new landing flaps. The WP-13F is used for propulsion.

The equipment is improved compared to the previous classic J-7 versions, but the most important sensor remains unchanged with the Type 226 radar rangefinder. Numerous components of the simultaneously developed J-7IIIA are used, such as JD-3 TACAN, 930-4 RWR and 941-4C chaff and flare dispenser. Other new devices include the Chinese JT-1 HUD, which initially proved to be very error prone, the SRT-651C radio station, the 8430 air data computer, the WL-7A radio compass, the 265 radar altimeter and a radio beacon receiver of unknown type.

The J-7E has an increased weapons payload of 1,600kg on four wing pylons. It can carry PL-2, PL-5B, PL-7 and PL-8 AAMs, HF-16 or HF-7 rocket pods, or 250kg or 500kg bombs. Unlike all previous versions, the guided missiles can now also be mounted on the outer wing pylons. Additional 480-litre drop-tanks can

A J-7IIIA on approach. As the landing flaps are only extended to 25 degrees, the boundary layer control system (SPS in Russian), which blows the landing flaps with bleed air from the engine, is not yet active, as it is only activated at deflections higher than 30 degrees. However, it cannot be deduced from the available sources whether this system was installed in the J-7III/IIIA and actually used in operation. (dlsunshy)

be mounted on the outer rails and a 720-litre drop-tank on the fuselage hardpoint.

The left-hand of the two Type 30-1 cannons, each with 60 rounds, has been removed, reducing the overall weight by 150kg. In place of the cannon, the original fuel tank of the MiG-21F-13 has been reinstalled, increasing the fuel capacity by 100 litres. The standard gravity refuelling of the MiG-21 is replaced by pressure refuelling, which reduces the refuelling time to six minutes. The hydraulic system has been redesigned to improve maintainability. The brakes were converted from compressed air to hydraulic actuation and equipped with electronic ABS and carbon composite brake discs, whereby the nose wheel brake was omitted. Other distinguishing features compared to earlier versions include the encapsulated angle-of-attack sensor on the nose and the rudder with powered auxiliary rudder. According to available photographic evidence, both the HTY-2 and the HTY-6 ejection seat is used in the J-7E. It is unclear whether the change was made during series production or as a retrofit. Like the J-7HH, the J-7EH version for naval pilots feature reinforced anti-corrosion measures for the airframe and engine.

Later modernised to the J-7L, the J-7E has the radar sight of the J-7G (the L in the version designation stands for radar sight) as well as components of the J-10. Externally, these aircraft differ from the J-7E by new antennas behind and under the cockpit and on the vertical tail.

J-7EB

The J-7EB is the variant of the J-7E for the 1st August Aerobatic Team of the Air Force, where 'B' indicates 'Demonstration'. This version has the 30-1 cannon on the right-hand side removed together with the ammunition feed, the wing pylons, the entire fire control system and other weapons-related components. The changed centre of gravity is compensated for by ballast and a smoke generator is installed under the fuselage. According to Chinese sources, the HTY-6 ejection seat is only fitted to the aircraft in the blue and white livery, while the HTY-2 is used in the red and white J-7EBs.

F-7MG

The F-7MG was created for export by equipping the J-7E with Western avionics. The core elements are the Super Skyranger ranging radar and the Type 956 combined HUDWAC. The avionics suite, consists of the KTR-908 and KTR-909 radio stations, the KNR-634A navigation system, the KTU-709 TACAN, the KDF-806 radio compass and the associated digital displays, came from AlliedSignal (now Honeywell) subsidiary BendixKing. Other avionics components include a Chinese electronic warfare system, a RWR, chaff and flare dispensers and colour displays. The cockpit is designed according to the HOTAS (hands and throttle on stick) principle and is equipped with the HTY-2 ejection seat. In the course of development, the F-7MG prototypes were retrofitted with the new one-piece windscreen.

F-7PG

The F-7PG has the double-delta wing of the F-7MG and is powered by the WP-13FI. Fuel capacity and maximum range are increased compared to the previously procured F-7P. The core component of the avionics is the Italian FIAR Grifo-7 PG multifunction radar. The electronic countermeasures system consists of the Chinese ARW9101 RWR, coupled with the chaff and flare dispenser.

Like the F-7MG and unlike the J-7E, the F-7PG has the second 30mm cannon on the left side of the fuselage and a one-piece windscreen. Other changes include a redesigned cockpit with ergonomic HOTAS instrument layout, oxygen generation system, global positioning (GPS) navigation system, external load management system, voice warning system and an electronic flight instrument system (EFIS) with two colour displays as well as radio compass, VOR/ILS and TACAN. The Martin-Baker Mk.10L ejection seat is used.

Guided missiles and unguided rockets as well as bombs from Pakistani, Chinese, US and French production can be carried on four multi-purpose racks under the wings. The missile armament also includes the Chinese PL-5E and PL-9C infrared AAMs, although various Sidewinder models can usually be seen on the aircraft in daily use.

J-7G

The J-7G is the first and only true all-weather, multi-role, combat aircraft among the J-7 versions for domestic use. Externally, it differs from the original J-7E model primarily in the one-piece windscreen and the modified antenna arrangement; in terms of equipment, it represents a qualitative leap forward. The new windscreen is also the prerequisite for the use of a helmet display. In combination with the new JL-10G pulse-Doppler radar, the cockpit designed according to the HOTAS principle and the PL-9C AAM, more effective combat management is possible.

Integrated satellite navigation and a precision bombing regime allow the use of retarded and guided bombs. In addition, the aircraft is equipped with an improved Type III IFF system, the LJ-2

On the J-7EB display version, the cannons were removed and the cannon fairings were closed. The pod under the fuselage was not an additional fuel tank but contained coloured oil for smoke generation. (Holger Müller)

The one-piece windscreen, which would go into series production with the F-7PG, dramatically improved visibility and allowed the use of a helmet display. Thanks to the unbraced design, the HUD is clearly visible. While the F-7MG prototype shown here still had the old HTY-2 ejection seat, the export versions were equipped with the Martin-Baker Mk.10L or the very similar Chinese HTY-6. (Holger Müller)

omnidirectional RWR and a chaff and flare dispenser. A datalink system is used to exchange information and commands with early warning aircraft. The HTY-6 zero-zero ejection seat is fitted as standard, as was the WP-13FI engine.

J-7GB

In the J-7GB demonstration version, which replaced the J-7EB, all armament components are removed, equipment components adapted and formation flight lights and a smoke control box installed. At the end of their deployment with the 1st August Team, the J-7GBs were refitted to the G standard and handed over to operational regiments.

F-7NM/F-7BG/F-7NI/F-7GS/F-7TN

The NM, BG, NI, GS and TN export models based on the J-7G have the double-delta wing, a WP-13FI engine and one-piece windscreen, but differ in the customised radar and other avionics equipment. The Chinese SY-80 (JL-10G) or KLJ-6E models with very similar performance data are specified as optional radar equipment.

Only for the F-7GS, which corresponds to the J-7G in essential aspects, there is more-detailed information about the equipment available. This includes a HUD with external load management system, coupled with the Type 8430 air data computer, voice warning system, colour video recorder, cockpit lighting compatible with night vision goggles and helmet display, a more precise angle-of-attack sensor, GPS and inertial navigation system, AD3400 multiband radio station, Type 605A IFF, KLJ-6E pulse-Doppler radar, WL-7 radio compass, 0101 HR A2 radar altimeter, LTC-2 artificial horizon, XS-6 radio beacon receiver, VOR, DME, ILS and TACAN. The cockpit is equipped according to the HOTAS principle.

Armament consists of one (F-7GS/F-7TN) or two (F-7NM/F-7BG/F-7NI) 30mm 30-1 cannons. The four underwing pylons can be loaded with a variety of air-to-air missiles, unguided rockets or bombs.

Together with the double-delta wing, newly designed wing pylons with a stepped shape were also being introduced. A mock-up of a PL-7 air-to-air missile can be seen on the inner pylon, while a 490-litre drop-tank hangs from the outer one. (Holger Müller)

The F-7BGs were delivered in this unique livery in various shades of blue, which were not used for any other customer. Due to the completely unpressurised hydraulic system, both the front and rear airbrakes are extended on this aircraft. (Holger Müller)

The F-7GS (and possibly others) has a Chinese helmet-mounted sight, which is used for target designation for the PL-9 AAM. The HTY-6 ejection seat is installed in all versions mentioned. Like the earlier F-7MBs, the F-7BGs are capable of towing an aerial target.

F-7BG-I

The final J-7 variant for export to Bangladesh is again based on the J-7G, but unlike the previous versions described, it represents a further development. The aircraft has significantly larger displays than its predecessor, the F-7BG, a multifunctional HUD, a cockpit layout based on the HOTAS principle, a helmet display and, according to Bangladeshi documents, is equipped with the SY-80 (JL-10G) radar. The armament comprises only a 30mm cannon on the right side of the fuselage and, among other things, the PL-5EII and PL-9C AAMs, LS-6 unguided and satellite-guided bombs and – according to as yet unconfirmed information – anti-ship missiles.

Twin-Seaters

JJ-7

The trainer version of the J-7 combines the front fuselage section of the Soviet two-seater with the rear of the J-7II, using neither the widened vertical stabiliser of the original MiG-21US nor its SPS boundary layer control system. To improve stability, two inclined fins replace the previously centred false keel under the fuselage. The two lateral airbrakes of the J-7II below the cockpit are replaced by a central one below the fuselage, as on the MiG-21US.

Both cockpit canopies open to the right and an HTY-2 ejection seat is installed in both positions. A cockpit intercom system enables

The cockpit of the F-7BG is dominated by a large, newly developed HUD. Below it are two relatively small multifunctional displays. (Holger Müller)

Landing gear test of an F-7BG. From this perspective, the unbraked front wheel of the J/F-7 versions with double delta wing is clearly visible. (Holger Müller)

The F-7GS's range of weapons, consisting of air-to-air and air-to-ground systems, clearly shows that the latest generation of J/F-7 has matured into a true multi-role combat aircraft. (Holger Müller)

F-7BG-I in overflight. From this perspective, the essential features of the last J/F-7 single-seat version – double-delta wings, one-piece windscreen, only one 30mm cannon on the left and the antenna with 'roof' behind the cockpit – are clearly visible. (Holger Müller)

When comparing the cockpit of the F-7BG-I with that of the predecessor model, the significantly larger multifunctional screens in the centre catch the eye. Overall, the arrangement of displays appears tidier and more ergonomic. (Holger Müller)

communication between student and instructor. All instruments and controls are available in both positions. Aircraft and engine controls, brakes and other systems are equipped with a disconnect mechanism so that the instructor can disconnect the student's controls and take over these himself. The two-seater is also equipped with an error simulation system. Red light illumination enables night flight operations. The avionics equipment includes the HK-03 optical sight, the CT-3 radio station, the WL-7 radio compass, the 262 radar altimeter and the XS-6B radio beacon receiver.

The enlarged fuel tank in the rear of the fuselage compensates for the reduction in fuel volume due to the enlarged cockpit. Like the Soviet trainers, the JJ-7 does not have a built-in cannon. Compared to the J-7II, the main landing gear has reinforced shock absorbers and tubeless tyres. The pitot tube is moved to the top right of the nose. The WP-7B (BM batch) engine is used. A 480-litre or 720-litre drop-tank can be carried on the fuselage hardpoint and guided air-to-air missiles, and unguided air-to-ground rockets and bombs on the wing pylons.

FT-7/FT-7B/FT-7BI/FT-7BB/FT-7BS/FT-7BK/FT-7BZ

Like the JJ-7, early export versions based on the JJ-7 had the HTY-2 ejection seat and also largely corresponded to the original type in terms of other equipment. The equipment is simplified and meets the needs of the generally financially weak recipient countries. These include the versions for Iraq (FT-7BI), Bangladesh (FT-7BB), Myanmar (FT-7BK), Zimbabwe (FT-7BZ) and Sri Lanka (FT-7BS).

Only the aircraft for Iraq and Zimbabwe – like the FT-7 prototype – are equipped with an electrically extendable periscope to improve the view of the instructor sitting at the rear. The retraction and extension of the periscope, which gives the instructor a view of the runway up to approximately 25 metres in front of the nose of the aircraft, is linked to that of the landing gear.

JJ-7A

The advanced JJ-7A training aircraft differs from its predecessors primarily in its avionics. Among other things, a HUD is installed as used in the J-7E. The aircraft has a video recording system and the FJ-1 flight data recorder, a new radio antenna has been added to the rear of the fuselage and the cockpit layout is changed. Air conditioning and the engine ignition have been improved and the oxygen supply system has been modified. The JJ-7A can carry a

Like the FT-7 prototype, the first exported two-seaters (here one of the FT-7BZ for Zimbabwe) have a periscope for the flight instructor. (Internet)

The later trainer versions for export, such as the FT-7BB for Bangladesh, are delivered without a periscope. Ejection seats from Martin-Baker replace the HTY-2 after a few years in service. (Holger Müller)

gun pod with a double-barrelled 23mm Type 23-3 cannon on the fuselage hardpoint. The fuselage structure, main landing gear and avionics have been revised with regard to maintainability, reliability and service life.

Initially equipped with the HTY-2 ejection seat, the JJ-7A later received the HTY-6, both during production and in the form of retrofits. The appearance of identical JJ-7A construction numbers, one beginning with A and the other with L, suggests the existence of a modernised JJ-7L similar to the J-7L. These aircraft do not have the central display in the rear cockpit, so they may also be modernised JJ-7s.

FT-7P

In the FT-7P two-seater version, tailored to the needs of the Pakistan Air Force, the fuselage has been lengthened by 0.61 metres with a section inserted behind the cockpit and the fuel capacity has been increased by 350 litres to 2,800 litres. A Type 23-3 cannon is installed under the fuselage. Like the single-seaters, the wings of the FT-7P have two wing pylons, the outer one of which are suitable for carrying additional drop-tanks. This brings the fuel quantity and range into line with those of single-seaters.

The FT-7P has the same operational capabilities as the single-seaters as it is the first MiG-21 twin-seater to be fitted with a radar – the same Grifo-7 as on the F-7P – and the HUDWAC. The installation of the omnidirectional LJ-2 RWR and a chaff and flare dispenser improves survivability, while modern radios and TACAN enhanced communication and navigation. A video recording system logs both flying activities and the use of weapons. The cockpit layout corresponds to that of the F-7P. The rear seat position is slightly raised to provide better visibility for the instructor. WP-7B and WP-7C engines are used.

JJ-7A with retrofitted or standard HTY-6 ejection seats, easily recognisable by the canopy breakers protruding far forward. Also clearly visible are the frame of the HUD in the front cockpit and the display in the centre console of the rear cockpit. (dlsunshy)

From this perspective, the significantly longer fuselage of the FT-7P compared to the classic FT-7 immediately catches the eye. But the two generations of the FT-7 can also be easily distinguished from other angles by the fairing of the wing root: on the JJ/FT-7 this starts below the front cockpit and directly above the auxiliary air intakes, but on the FT-7P it only starts between the two cockpits and about half a metre behind these openings. (Holger Müller)

FT-7N/FT-7PG/FT-7BG/FT-7NG/FT-7BG-I

The two-seater export versions delivered from 1992 onwards are based on the stretched FT-7P. This applied both to the FT-7N for Iran, which still corresponds to the JJ-7 in terms of equipment, and to the following two-seaters for Pakistan, Bangladesh, Namibia, Nigeria and Tanzania, each of which are equipped with avionics corresponding to the F-7MG/J-7G-based single-seaters delivered at the same time. All have the Type 23-3 cannon under the fuselage and user-specific adaptations to the equipment.

JL-9/JJ-7B/FTC-2000/FTC-2000G

A key feature of the JL-9 and FTC-2000 is the completely redesigned front fuselage. Lateral air intakes and a fixed nose create space for the installation of a multifunction radar.

The JL-9 has been given a double-delta wing similar to that of the J-7E, but the sweep of the outer wing was reduced to 37 degrees. The landing flaps and ailerons of the delta wing J-7 are retained, so that the trailing edge of the wing is straight as on the JJ-7. Increased wing area and wing aspect ratio reduce the induced drag. Lift, sub- and transonic performance at low and medium altitudes and range have been improved. Take-off and landing behaviour are significantly better than with the classic J-7. The outwardly inclined air inlets at the bottom side of the fuselage ensure the air supply for the WP-13F(C) engine even at high angles of attack.

Thanks to the side air intakes, the airflow to supply the engine is no longer channelled through the nose and under the cockpit. This allows the nose of the aircraft to be much flatter and the front seat to be positioned lower. Together with the redesigned cockpit glazing with an unstrutted windscreen, a 14-degree inclination of the front cockpit roof and a five-degree inclination of the rear cockpit roof, this results in a significant improvement in visibility for both pilots. The cockpit layout corresponds to that of newer generation fighter aircraft, so that student pilots are familiarised with the use of head-up and multifunction displays during training. The front cockpit has a camera that transmits pictures of both instruments and student to a display in the rear cockpit. The avionics are Chinese-made and consist of a glass cockpit, an air data computer, a JL-10G pulse-Doppler radar, an inertial and a GPS navigation system, a weapon station management system and a mission computer.

The JL-9 and FTC-2000 as well as the JL-9G and FTC-2000G differ in numerous externally visible features. While the rear fuselage of the original version, including the vertical stabiliser, the brake chute container and the two fins below, corresponds to that of the JJ-7/FT-7, the G version has a completely new, larger vertical tail with a

The FT-7N for Iran combines the airframe of the FT-7P with the (purely Chinese) equipment of the JJ-7, including the HTY-2 ejection seats. (Holger Müller)

While at least the front cockpit (left) of the FT-7 is equipped with a HUD, the instructor's seat (right) has only minimal equipment. (Holger Müller)

Like the single-seater F-7BG, the FT-7BG was delivered in this striking blue livery. The slightly raised seating position of the flight instructor, which was intended to improve visibility from the rear seat, is clearly recognisable. The proximity to the canopy requires shorter canopy breakers for the rear seat. (Holger Müller)

The final development stage of the two-seater can be recognised externally by the modified antennas, such as the model with the 'roof' on the back of the fuselage and the flat blade antennas of the VOR receiver on both sides of the vertical tail. (Holger Müller)

In the FT-7BG, the flight instructor also has a display – located at the top centre of the central instrument panel. It is unclear whether this rather improvised device is a multifunctional display for flight information or just a camera display. (Holger Müller)

On the FT-7BG-I the rear cockpit display has been modified again. It is interesting to note that the fan, which was already used on the MiG-21 two-seaters to improve cockpit air conditioning, is still used here. (Holger Müller)

straight leading edge and a rearward sloping upper end. The brake chute container and two fins below have been omitted. Only the stabilisers (elevators) still have the classic MiG-21 shape. Equally striking are the differences in the air intakes and wings. On the JL-9/FTC-2000, the leading edges of the air intakes run at right angles to the longitudinal axis of the aircraft and the air intakes are separated from the fuselage by diverters. The leading edges of the air intakes of the JL-9G/FTC-2000G are negatively swept and diverterless with a central bump.

The leading edges of the wings of the JL-9 and FTC-2000 are divided into two straight sections, with a sweep of 57 and 37 degrees, respectively. On the G version, the inner section is curved while the straight section with a 37-degree sweep remains on the outside. The new wing design reduces the maximum speed of the FTC-2000G, but at the same time the payload increased to 3,000kg.

Despite the modernised avionics, the JL-9 does not have a fly-by-wire control system but continues to use a mechanical control system. The HTY-6 ejection seats are equipped with canopy breakers, which shorten the ejection process and thus improve safety when ejecting at low altitudes.

A dummy air refuelling probe can be mounted on the nose. Unlike the previous export two-seaters, the JL-9(G) and FTC-2000(G) do not have a built-in cannon. However, it is possible to mount a pod with a Type 23-3 cannon under the fuselage. One fuselage and four wing pylons allow air-to-air and air-to-ground armament as well as additional drop-tanks to be carried.

The FT-7BG-I differs externally from the FT-7BG except for the – now harmonised – paint scheme with the VOR blade antennas on the vertical tail. The 2704 seen here with the construction number (T)P2104 was probably the last F-7 export two-seater to be produced. (Holger Müller)

A new, centrally positioned display and a smaller model on the left-hand side have found their way into the rear cockpit of the FT-7NI. While the smaller device is obviously a multifunctional display, the purpose of the larger one remains unclear. The lack of controls and adjustment options only for brightness and contrast suggest that it is only a camera display. (Holger Müller)

Unlike the single-seaters, the final two-seater generation has the same gravity refuelling system as the MiG-21 filled via the central fuel filler neck on the rear of the fuselage. (Holger Müller)

To improve visibility for the flight instructor, the student pilot's seat is positioned significantly lower in the FTC-2000. This is made possible by the flatter nose and the lateral air intakes. In addition, the unbraced windscreen and inclined cockpit roofs contribute to better visibility. (AVIC)

10 ENGINES

Three generations of jet engines power the various versions of the J/F-7. While the WP-7 is an unchanged copy of the Soviet R11F-300, the subsequent WP-7B incorporates numerous Chinese detail solutions. The final WP-13 is based on the Soviet R13-300, combined with solutions from the WP-7B and other developments.

WP-7

The WP-7 is a twin-spool turbojet consisting of an axial compressor, tubular ring combustion chamber, turbine and afterburner with adjustable thrust nozzle.

The low-pressure compressor and turbine form the low-pressure rotor (LPR), while the high-pressure compressor and turbine form the high-pressure rotor (HPR). Both rotors are only gas-dynamically coupled, which ensures stable operation of the compressor even in regimes outside the design range. Apart from the inlet section and thrust nozzle, the design as a twin-spool engine makes it possible to dispense with variable geometries and to keep the control system comparatively simply.

The engine operates in conjunction with a supersonic air intake section, which – like the outer contour of the thrust nozzle – is structurally part of the airframe. To ensure operation, there are subsystems for lubrication, cooling, ventilation, power and fuel supply, control, drainage, starting, monitoring and fire extinguishing. An accessory carrier attached to the engine is used to accommodate the power generators for supplying engine and equipment as well as for starting.

The air intake section contributes significantly to the stable and effective operation of the engine. The air flow is controlled by three basic ways to vary geometry. The cone is moved axially depending on the compressor pressure ratio and stabiliser deflection. At high Mach numbers, the antisurge doors open when the engine is throttled; at low pressure in the inlet section, the auxiliary air intakes open.

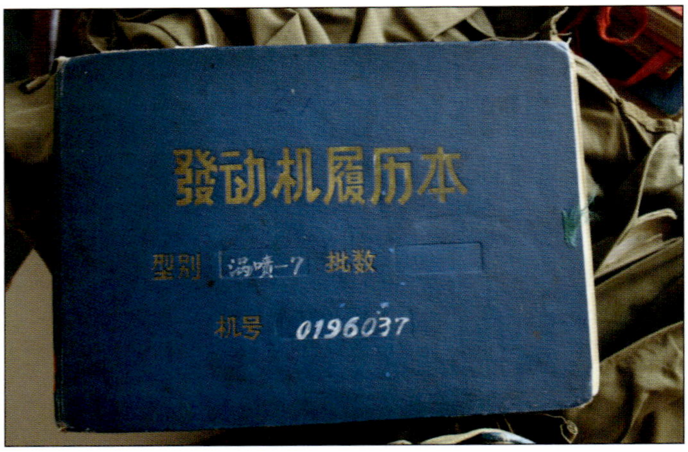

In the beginning, copying of Soviet technology by the Chinese also included the corresponding documentation. As a result, the engine logbook of a WP-7 looked exactly like the corresponding document for an R11 – apart from the Chinese characters. (Stefan Büttner)

The compressor of the WP-7 is divided into two cascades of three stages each, with each of these cascades being driven by its own turbine. The first three stages of the compressor form the low-pressure compressor, which is driven by the second turbine stage, the low-pressure turbine, while the first turbine stage, the high-pressure turbine, drives the last three compressor stages of the high-pressure compressor.

The WP-7 is equipped with a tube-ring combustion chamber consisting of 10 flame tubes. The flame tube heads are connected to each other in a ring by ignition nozzles. The combustion chamber is ignited by two starter ignition devices between two flame tubes. The main mass, the secondary air, which accounts for approximately 70 percent of the air flow, enters the flame tubes through several rows of openings, mixes with the hot combustion gases, cools them and ensures a temperature distribution that does not impair the strength of the turbine blades under all conditions.

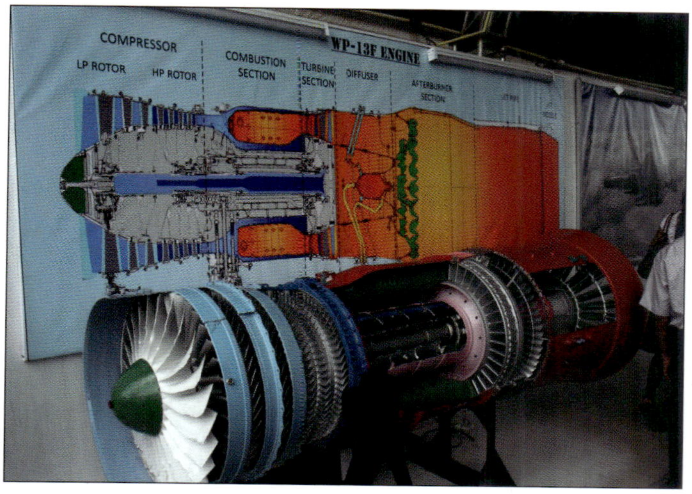

Like the R11, the WP-7 – shown in the cutaway model in the foreground – has six compressor stages, while the WP-13 pictured behind it has eight, like the R13 it is based on. (Holger Müller)

Each stage of the two-stage turbine consists of a stator and a rotor. The turbine is cooled using air extracted from the combustion chamber and compressor. To increase the thrust without increasing the air flow rate, the engine is equipped with an afterburner behind the turbine.

The energy required to operate the aircraft's systems is taken from the engine's shaft system in mechanical form and converted into electrical or hydraulic energy. The devices required for this are mounted on the engine's accessory carrier and are driven by the high-pressure rotor. This includes the main fuel pump, the afterburner fuel pump, two hydraulic pumps, the pre-pressure fuel pump, the lubricant unit, the centrifugal air separator and the DC starter-generator.

The engine is started by an electric starter, which operates as a DC generator for the on-board power supply once the starting process is complete. There is also a starter system with gasoline, which is fed to the starter blocks by an electrically operated starter pump.

The engine's autonomous lubricant system realises pressure circulation lubrication. The fuel flows from the extraction tank to the engine under the influence of gravity, low-pressure ventilation and the extraction pump. The extraction pump is installed in an inverted-flight tank for short-term fuel delivery even at negative g-loads. The fuel is transferred from the extraction pump to the booster pump, which increases the fuel pressure and dampens pressure variations. As the fuel flows through the fuel-lubricant unit, the fuel cools the lubricant and heats up in the process. The engine is designed for operation with different types of kerosene.

The electromechanical fuel control system ensures stable operation of the engine in all flight and engine regimes. The engine is controlled mainly by changing the amount of fuel in the main combustion chamber via the main fuel pump. The afterburner is controlled by the afterburner pump in conjunction with the electrohydraulic thrust nozzle control system.

The operating time between two overhauls and the overall service life of the original WP-7 is extremely low at 100 and 300 hours respectively due to numerous design and production deficiencies.

WP-7B

The WP-7B is based on the WP-7 and differs from its predecessor in that it has a reinforced first compressor stage, hollow, cooled turbine blades and a differently designed afterburner. The 100 K increase in turbine inlet temperature enabled by the hollow blades increases both maximum and afterburner thrust and reduces fuel consumption by two percent.

The star-shaped flame stabiliser of the WP-7 resembles that of the R11F-300 and is retained for the later WP-7B. Differences appeared on subsequent variants only. (Holger Müller)

In the WP-7B (batch M), the gasoline starter of the original WP-7B (batch 02) is replaced by a kerosene starter, which simplifies maintenance and use by eliminating the gasoline tank. In the subsequent WP-7B (Lot BM), improved design reduced the weight of the engine by 17 kilograms.

Since the PL-7 air-to-air missile frequently causes engine flameouts due to its high smoke emission, the WP-7B (Lot BM) is equipped with a device for restarting in the air. An ignition device coupled to the launch control system ensures that the engine is constantly re-ignited.

The operating time of the WP-7B until the first overhaul is increased from 100 to 200 and finally to 300 hours in the course of production. The total service life increased from 300 to 600 and finally to 900 hours at the end of production in 1992.

WP-7C

Compared to the WP-7B the WP-7C has improved thrust characteristics and increased corrosion resistance. The flame stabilisers of the afterburner are no longer arranged in a star shape but now in a ring.

WP-13

The WP-13 is based on the Soviet R13-300, but – at least in later development stages – also incorporates elements from the development of the WP-7B with regard to the compressor, combustion chamber, turbine and afterburner.

As with the MiG-21, the rear fuselage section of the J/F-7 has to be removed to replace the engine. To fix the engine with the tail section removed, a special auxiliary frame is mounted at the separation frame, which also allows the engine to be tested without the tail. (Holger Müller)

Design differences compared to the WP-7 include revised, reinforced and additional components and devices, improved fuel supply, cooling and control as well as new materials, primarily titanium alloys. The twin-spool principle and external dimensions of the engine are retained.

By extending the high-pressure compressor by two additional stages to a total of five stages with the same pressure ratio, the compressor stages are relieved and the stability reserve and efficiency are increased. A ribbed inlet ring also contributes to improved stability. The combustion chamber has a modified flame tube design. Compared to the R11/WP-7, the newly designed turbine increases the turbine inlet temperature and thus the efficiency, improves cooling, thermal and mechanical strength and improves the energy transfer from the turbine to the compressor and thus the compressor characteristics. The afterburner chamber and thrust nozzle have been reinforced in view of the increased heat and pressure load.

The new pre-pressure fuel pump has an extraction nozzle for additional afterburner supply and a system for reducing pressure oscillations at high altitudes. The fuel pump, which is also new, has an overspeed mechanism to increase the speed in the second afterburner stage and a barometric altitude corrector.

As a result of these measures, the thrust increases compared to the WP-7 across the entire power range. A second afterburner stage with further thrust increase is available above Mach 1.5. Increased afterburner chamber pressure as well as higher air throughput increase efficiency. Like the WP-7B, the WP-13 has a device for preventive restarting when the combat button is pressed.

In the further development WP-13AII, the housing of the 2nd to 7th compressor stage is made of cast titanium instead of steel. Hollow, air-cooled blades in the first turbine stage allow a higher turbine inlet temperature and thus an increase in thrust. At the same time, the weight is reduced by almost 20 kilograms.

WP-7B, installed in F-7P and FT-7P. (Stefan Büttner)

The engine of an FT-7BB shown here has ring-shaped flame stabilisers, which are also typical of the Soviet R25. Obviously, this is the WP-7C. (Holger Müller)

The WP-13 versions F and FI feature design elements for which there are no equivalents in the Soviet R11/13/25 engines. These include the star-and-ring-shaped flame stabiliser of the afterburner. (Holger Müller)

Based on the WP-13AII, the WP-13F has been improved in numerous aspects, particularly the structure and materials of the turbine and afterburner. Maximum thrust and afterburner thrust are significantly increased compared to the WP-7. Fuel consumption has also been reduced and reliability and maintainability improved. The flame stabiliser of the afterburner has been redesigned and now consists of star-shaped and ring-shaped components.

With the final WP-13FI, maximum and afterburner thrust can be slightly increased again. The engine has a revised first compressor stage with the number of blades reduced from 24 to 19, similar to the Soviet R25, and an increased mass flow rate. The compressor blades have boundary layer suction. The materials, processing and heat protection coating of the parts with high thermal loads are improved so that the turbine inlet temperature reaches 1,015 degrees Celsius, thus increasing the thrust.

The operating time between two overhauls and the total service life for all WP-13 variants are 300 and 900 hours respectively.

WP-14 Kunlun

This engine, developed by the 606 Institute since 1984, is officially the first turbojet engine to be developed entirely in China, but 70 percent of its parts are identical to those of the WP-13F. Ground tests began in 1993, but it was not until 2002 that it received state certification and flight testing began in 2004. In terms of afterburner thrust, the WP-14 is the most powerful engine installed in a any MiG-21 derivative. Only the R25 in so-called 'special regime' has more thrust. The further increase in performance compared to its predecessor, the WP-13, is achieved by design improvements without radical changes. This engine does not have a digital control system, but an electromechanical fuel control system like all previous models. The overall service life, which is now 1,500 hours, and the operating time until the next overhaul, which is 500 hours, have been significantly increased. In 2007, a WP-14 was installed in a J-7G on a trial basis. However, for unknown reasons, it is not installed as standard in the J-7. All subsequently delivered aircraft and also the JL-9/FTC-2000 use the WP-13FI.

Table 1: Engine Characteristics

Engine		**WP-7**	**WP-7B**	**WP-13**	**WP-13F**	**WP-13FI**
Aircraft version		**J-7/7I**	**J-7II/ F-7A/B/M**	**J-7C**	**J-7E**	**J-7D/G**
Compressor stages LPR + HPR		3 + 3	3 + 3	3 + 5	3 + 5	3 + 5
Maximum static thrust	kN	38.3	42.3	39.2 40.2	43.2 44.1	45.1 44.2
Maximum static thrust AB	kN	56.4	59.8	62.8 64.7	63.3 64.7	66.7 65.4
Spec. fuel consumption at maximum static thrust	kg/kN x h	98.9	103.0	97.9	100.9	100.9
Spec. fuel consumption at maximum static thrust AB	kg/kN x h	234.0	204.0	229.0	209.0	209.0
Maximum compression		8.9	8.9	8.8	8.8	9.2
Turbine inlet temperature	°C	915	1,015	950	1,015	1,015
Air flow rate	kg/s	64	65	66–67	66–67	68–69

11 EQUIPMENT AND AVIONICS

The equipment of the J/F-7 was initially based entirely on Soviet components, which were gradually replaced by Western technology or Chinese developments based on it. Due to the limited installation space available – especially for the radar rangefinder and the radar – the equipment of the J/F-7 remains comparatively weak across all versions. For many Chinese components, especially of the early J/F-7 versions, the designations are not known. In this case, the names of the corresponding Soviet equipment are used.

Radars and Radar Rangefinders

The radar is the primary sensor of every modern fighter aircraft and thus essentially determines its combat characteristics. Most versions of the J/F-7 are only equipped with a radar rangefinder and not with a full-featured radar, which massively limits their capabilities in modern air combat. Only the versions for Pakistan, the J-7III/IIIA, the J-7G family and the retrofitted J-7L have a radar in the true sense of the word.

As the original Soviet avionics proved to be inefficient and the Chinese copies were even weaker, China orientated itself on Western technology in its further development. For example, unlike airframe and engine, the JL-7 radar of the J-7III was not based on the corresponding Soviet component, but on a captured US device. Later, purchased Western systems or their copies were installed, such as the GEC-Marconi Skyranger radar rangefinder and its Chinese version Type 226 as well as the Grifo radar of the F-7P/PG. Only in J-7G variants Chinese radars can be found, but also with Western genes.

CL-2 (SRD-5MK Kvant)

The CL-2 radar rangefinder, based on the Soviet SRD-5MK Kvant, measures the distance to an aerial target by measuring the transit time of the reflected transmitted signal and calculates the speed at which the target is approached. Alternatively, it can also be used to measure the distance to prominent ground targets. The radar rangefinder works with a fixed antenna in the air intake cone, so that the target must be within the radio lobe with a switchable angle of 18 degrees (operating mode 'wide beam') or six degrees (operating mode 'narrow beam') for detection. The maximum range is seven kilometres, that of Chinese CL-2 copy only three kilometres and that of advanced CL-2A eight kilometres. In the 'wide beam' mode, which is used for cannon and unguided rockets, the measurement range is between 300 and 3,000 metres, in the 'narrow beam' mode for the R-3S (K-13) missile between 800 and 7,000 metres.

The distance is displayed by the UD-1 analogue instrument and simultaneously a signal is fed into the optical sight to increase accuracy. In cooperation with the WRD-2A weapon computer, the permissible firing distance of the missiles is calculated. When the minimum firing distance is reached or undershot, this is signalled optically and acoustically. The CL-2 operates at a frequency of 9,370 MHz, a pulse frequency of 800 Hz and a pulse power of 5 to 7 kW. Its mass without cable is 25 kilograms. CL-2 and CL-2A are built into the domestic versions J-7 and J-7I.

CL-3 (Type 222)

The CL-3, developed by the China Leihua Electronic Technology Research Institute (607th Institute, CLETRI), is said to be the first radar rangefinder developed in China. The device, also known as Type 222, has a longer range than the CL-2 and is used in the J-7II as well as in the F-7A and F-7B export versions.

Type 226

There is contradictory information on the Type 226 device, also developed by the 607th Institute (CLETRI), which is the primary sensor for all single-seat versions up to the J-7E and the JJ-7 two-seater. According to some sources, this is a copy of the Skyranger (see below); others speak of an independent development. The first version is supported by the fact that the specified range of approximately 10 kilometres is identical to that of the Skyranger.

Skyranger 7/7M

The Skyranger 7 radar rangefinder developed by GEC-Marconi allows the detection of aerial targets up to a distance of almost 10 kilometres. The device weighs 41 kilograms and has a parabolic antenna. As a result of the tests in the F-7M, GEC-Marconi equipped the device with technology to suppress ground clutter, creating the Skyranger 7M. The Skyranger is installed in the J-7IIA and the Skyranger 7M in the F-7MP/P.

JL-7/JL-7A

The larger diameter of the inlet cone of the J-7III offered the Chinese aviation industry the opportunity to install a real radar in a J-7 for the first time. Previously, the available space and lack of know-how in the miniaturisation of electronic assemblies had prevented this. Starting point for development of the JL-7 was not the Soviet RP-21 of the MiG-21MF or the more powerful RP-22 of the MiG-23MS (which China also received from Egypt), but an indigenous system with US roots.

The development history begins as early as 1966, when China came into possession of the NASAAR R-14A radar from a US F-105 Thunderchief. Based on this, types 317 and 317A were developed, the latter of which formed the basis of the JL-7. Due to its derivation from a fighter-bomber radar, the device, which was developed by the 607th Institute (CLETRI) from 1980 and certified in 1986, has both air-to-air and air-to-ground functions.

The JL-7 is a mono-pulse radar with a wavelength of 2cm and a maximum range of 28 to 30 kilometres in detection mode and 15 kilometres in target tracking mode at an operating altitude of above 800 metres. It weighs 115 kilograms and the average operating time between failures is between 50 and 70 hours according to various sources. The device has an integrated IFF, electronic counter measures and interfaces to a ground test system and can operate in search, intercept, automatic tracking and air-to-ground modes.

The external dimensions and the installation volume of 230 litres prevent it from being used in the J-7s with the narrow air intake, so that only a few JL-7 units were produced. The JL-7A radar of the J-7IIIA, which was also only produced in small series, is said to have a look-down-shoot-down capability.

Super Skyranger

After the positive experience with British avionics in the F-7M family, GEC-Marconi was selected in 1995 as the supplier for the radar of the F-7MG in 1995. However, the purpose-designed pulse Doppler radar Super Skyranger (SSR) was not convincing and development was stopped after prototype trials in 1998. Only the F-7MG, which was also only a prototype was equipped with it.

The SSR has two air-to-air operating modes for the use of missiles or cannons respectively and a maximum range of 15 kilometres in the first mode and five kilometres in the second. The minimum range in these modes is 300 and 150 metres respectively. There is also an air-to-ground mode available. The radar has a planar slotted array antenna and operates in the frequency range of around 10 GHz with a maximum beam deflection of ± 30 degrees, but this is limited to ± 20 degrees due to the installation situation in the F-7MG. The operating modes include search within the field of view of the HUD, vertical search, fixed beam, target tracking and range finding – the latter also during the search. The Super Skyranger has a look-down-shoot-down capability and is connected to the on-board computer via an ARINC 429 or MIL-STD-1553 data bus. It can track up to eight targets simultaneously and engage one of them.

Grifo-7

As the Chinese aviation industry was not yet able to supply a radar for the F-7 variants with the narrow air intake at the beginning of the 1990s, Pakistani requirements could only be met by procuring Western technology. The FIAR Grifo pulse Doppler radar selected by Pakistan to equip the F-7MP/P was originally developed in the early 1990s to equip modernised F-5s, but has since been adapted to numerous aircraft types with very limited installation space by fitting them with differently shaped antennas. The radar operates in the frequency range around 10 GHz and has operating modes for search, target tracking, air combat and engagement of ground targets, whereby up to eight targets can be tracked in air-to-air mode (according to other sources only four, one of which can be engaged).

The variants for the F-7, known as Grifo-7 or Grifo-7PG (with a swivelling antenna with a higher horizontal detection range), weigh only 55 kilograms due to the small antenna. The unit is cooled independently of the aircraft systems, which considerably simplifies installation. One of the differences between the two variants is the vertical and horizontal deflection of the antenna, which is limited to ±10 degrees for the Grifo-7 and ±30 degrees for the Grifo-7PG. The connection to the on-board systems, especially the head-up display, is made via the ARINC-429 bus as standard, optionally via the MIL-STD-1553B bus. The range is specified as approximately 55 kilometres.

Grifo radars are manufactured under licence by the Kamra Avionics, Electronics and Radar Factory of the Pakistan Aeronautical Complex (PAC). Installation of these radars makes the F-7P and F-7PG versions true all-weather fighters.

JL-10G (SY-80) and KLJ-6

Only for the last J/F-7 generation are Chinese-made radars available. After the use of the already proven and powerful Grifo was discarded for cost reasons, two models were developed in parallel and apparently both were fitted as standard. However, the information on this is patchy and contradictory. It is also unclear to what extent Western technology was available as a basis for development. That this is the case seems more than likely in view of the great technological leap that Chinese industry is making with these devices. Numerous sources point to Israel as the origin of this technology, which is plausible for several reasons. For example, there has been Chinese-Israeli cooperation in licensed production of the Python 3 air-to-air missile as the PL-8. In addition, Israel has developed high-performance radars and has been offering these specifically for the modernisation of the MiG-21 since the early 1990s. The fighter variant LanceR C of the modernised Romanian MiG-21 is equipped with the Israeli EL/M-2032 radar from IAI ELTA, the fighter-bomber and trainer variants LanceR A and B with the EL/M-2001 radar rangefinder from the same manufacturer. It is therefore quite logical that the technology of the EL/M-2032 was incorporated into the Chinese developments. At the same time, it is

FIAR presented its Grifo radar at Airshow China 1998. While the model shows the installation situation in the Northrop F-5, on rear left display panel the configuration for the F-7 can be seen. (Holger Müller)

In this advertisement, FIAR emphasises the Grifo's versatility in fighter aircraft with very limited installation space, such as the F-5, Mirage III, L-159 and the F-7. (FIAR)

unlikely that – as claimed by numerous sources – the EL/M-2001 or its Italian derivative, the Pointer 2500, form the basis of the Chinese radars, as these devices are, as already mentioned, radar rangefinders with significantly lower performance. However, even if the technology of the EL/M-2032 is used to a large extent, Chinese radars have to lag behind its performance in the MiG-21 – especially in terms of range, which is 80 kilometres there – as the narrow intake section only allows the installation of a much smaller antenna and makes heat dissipation and power supply difficult.

For the first time, two parallel developments – the JL-10G and KLJ-6 – compete for use in the final J-7 versions. The JL-10G was developed by the 607th Institute (CLETRI) and the KLJ-6 by the Nanjing Research Institute of Electronics Technology (14th Institute, NRIET). Between September 2002 and March 2003, both systems were subjected to comparative testing, in which the JL-10G finally prevailed. After further tests and successful certification, low rate initial production began in July 2004.

The JL-10G is based on earlier and already tested variants of this radar, such as the JL-10A used in the JH-7A, China's first pulse Doppler radar. Its development began as early as 1986. The JL-10G operates in the X-band with a medium pulse repetition frequency, offers operating modes for search, target tracking, air combat and engagement of ground targets and has functions for interference suppression and automatic target detection. The total weight is 60 kilograms, the installation volume less than 40 litres. It can be used at altitudes between 0 and 18,000 metres, with antenna swivel range of ±30 degrees. The detection range is 30 kilometres; individual targets can be tracked up to 26 kilometres. The average operating time between failures is 100 hours. The system includes a training device and intelligent diagnostic equipment.

The export designations of the JL-10G are SY-80 and SY-80A (improved version – SY is the abbreviation for Shen Ying, Celestial Eagle). To save weight, the radar does not have its own control unit. Its functions are displayed directly on the aircraft's head-up display. The JL-10G can work together with a helmet-mounted sight and can assign targets for the PL-8 and PL-9 air-to-air missiles, using their off-boresight capabilities.

Between 2004 and 2008, 200 JL-10Gs and SY-80s are said to have been produced for J-7G, including its export variants, to retrofit the J-7E under the designation J-7L and for installation in the JL-9. To date, the JL-10G is the only Chinese radar to be used in real combat. On 9 September 2008, a Sri Lankan J-7GS used its SY-80 to shoot down a Z-143 of the Tamil LTTE at an altitude of 500 metres.

The general description of the KLJ-6E is largely similar to that of the JL-10G or SY-80. It is also a pulse Doppler radar that operates in X-band and has air-to-air and air-to-ground functions. According to the illustrations provided, the antenna size and installation volume correspond to those of the JL-10G and the weight is also stated to be around 60 kilograms. There are very different specifications for the range (without further details as to which mode these apply to), ranging from 15 to 30 to 85 kilometres. If the latter figure is correct, it probably refers to the KLJ-6F with target designation capability for medium-range air-to-air missiles. However, this radar and the corresponding missiles are not available for the J-7, or at least have not been ordered by any customer. The KLJ-6 can track six targets and engage two simultaneously. The average operating time between failures is said to be 150 hours.

Various sources suggest that J-7s is indeed delivered with both systems – SY-80 and KLJ-6. So far, however, there is only evidence of the SY-80 being used in export versions – in addition to the aforementioned J-7GS, the Bangladeshi F-7BG-I is equipped with this radar, according to information provided to the author by the operator.

Other Soviet and Derived Chinese Systems

The avionics of the MiG-21F-13 and the early J/F-7 versions up to the J-7II and F-7B are – even for the time of their production – comparatively meagre. In addition to the aforementioned radar rangefinder, the equipment includes the ASP-5ND optical sight, the R-802 radio, the ARK-10 radio compass, the MRP-56P radio marker receiver, the RW-UM radio altimeter, the SOD-57M transponder, the SRO-2 friend or foe recogniser, the SPO-2 Sirena-2 radar warning receiver and the KAP-2 autopilot.

The ASP-5ND is a gyro-stabilised, semi-automatic optical sight and serves as a targeting device for the use of cannon, rocket and bomb armament. The device works alternatively in combination with the rangefinder CL-2/CL-2M or with manual range input.

The R-802 radio station operates in the VHF range (100 to 150 MHz) and is used to establish radio communications with all ground stations (flight control, fighter control centre, command post) and with other aircraft. The frequency is set electromechanically with a channel spacing of 83.33 kHz. The range is up to 250 kilometres at an altitude of 10,000 metres.

Radio navigation is carried out using the ARK-10 medium wave radio compass, which continuously determines the course angle to received MF transmitters and displays it on the course display instrument. That allows, among other things, to approach such a transmitter, determine position or determine and correct course. Thanks to coupling with a distance counter, the ARK-10 can permanently calculate and display the distance to a transmitter. A rotatable loop antenna, which is housed underneath the fuselage, is used as a direction-determining element.

The MRP-56P marker beacon receiver is used to signal the overflight of radio markers and also acts as a radio aid during the landing approach. As a rough marker of the landing path, it signals the overflight of the distant and near markers at the 4,000 and 1,000 metre points. The device's antenna is located at the bottom right of the rear fuselage.

The RW-UM radio altimeter measures and displays the height above ground in the altitude range 0 to 600 metres. At the same time, a so-called dangerous altitude can be pre-set in fixed steps. If the altitude falls below this level, the pilot is given an acoustic and visual signal/warning. This navigation aid is used both for safety during approach and low-level flight and as an aid for shooting at ground targets. The antennas of the radio altimeter are mounted below the wings.

The SOD-57M transponder (product 040) receives signals from radar stations in the decimetre range, converts them, amplifies them and transmits them again. This increases the range of the radar stations when detecting the aircraft, minimises interference and weather influences during detection and allows information on the aircraft's altitude and individual identification to be transmitted. The antennas of the SOD-57M are mounted on and to the side of the vertical stabiliser and on the leading edge of the wing.

The SRO-2 IFF (product 020 Chrome) is an active secondary emitter and co-operates with the ground-based radar stations (ground-to-air interrogation) and radars (air-to-air interrogation) in the Kremniy system. This makes it possible to identify own and enemy aircraft in interception process.

The SPO-2 Sirena-2 RWR warns the pilot optically and acoustically of impinging radiation and roughly indicates the direction of the radiation source.

The KAP-2 autopilot is used to automatically control and stabilise the aircraft, thereby increasing safety in all flight conditions and reducing the pilot's workload. As a single-channel autopilot, the KAP-2 acts exclusively on the aircraft's roll axis via the ailerons. It has operating modes for damping short-term oscillations and for maintaining predefined bank angles.

Other Western Systems and Indigenous Developments

Until the 1980s, the Chinese aviation industry was only able to equip aircraft with avionics that reached the performance of Soviet components from the time when licence production began, if at all. This only changed when the Chinese aviation industry gained access to Western technology through the procurement of British avionics in the early 1980s.

While an optical sight such as the ASP-5ND only displays target-relevant information, head-up displays provide the pilot with all flight and mission-relevant information in the field of vision. The Type 956 head-up display developed by GEC-Marconi (now BAE Systems) for the J/F-7 has an integrated weapon aiming computer, which is why the system is referred to as HUDWAC (Head-Up Display Weapon Aiming Computer). The system, which significantly increases the fighting power of the J/F-7, is installed in the J-7IIA, J-7IIM, J-7IIK, F-7M, F-7MG and F-7MP / P. Together with the HUDWAC, other British avionics components are supplied which form the basis for the subsequent Chinese developments, such as the Type 50-048-02 digital air data computer and the 2032 firing camera connected to the HUD.

F-7P simulator in Mianwali. The device is a fixed-base simulator used for instrument flight training. Above the central instrument panel is the HUD unit with integrated weapon aiming computer (HUDWAC) Type 956 from GEC-Marconi (now BAE Systems). (Holger Müller)

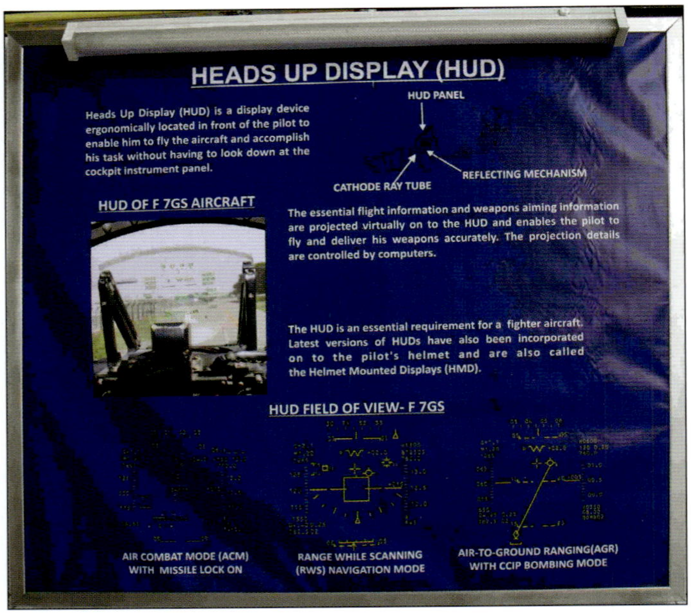

The display modes of J/F-7G's new HUD include an air combat mode, a navigation mode and an air-to-ground mode with calculation of the Constantly Computed Impact Point (CCIP). (Holger Müller)

Left: The brochure from Xian Xibei Optoelectronic Instruments clearly shows the difference between the model based on the British 956 HUD (right) and the Chinese development with two parallel screens (left) used from the J/F-7G onwards. (XBOE)

Even though a prototype of a Chinese HUD with an interface to a digital air data computer, developed by the Taiyuan Aeronautical Instruments Factory, was available as early as 1984 and was installed in a J-7II the following year, it can be assumed on the basis of the external similarity that the later Chinese devices for this purpose were at least based on the above-mentioned British ones. This applies to both the JT-1 HUD and the 8430 weapon sighting computer. Like the original model, the latter is capable of calculating both Constantly Computed Impact Points (CCIP) and Constantly Computed Release Points (CCRP) and displaying them in the HUD. This combination, in which an external load management system is also integrated, is used from the J-7E onwards. A new generation of HUD was not introduced until the J/F-7G generation, which differs from the original British model in external appearance, for example as it has two parallel screens.

The British AD-3400 multi-band radio station, which is used from the J-7IIA onwards and in the F-7M, F-7N and F-7GS, among others, represents a technical quantum leap thanks to digital technology, encryption (type AA3460) and a range of over 400 kilometres at an altitude of 1,200 metres. The F-7 modifications for Pakistan use the US radio stations Raytheon AN/ARC-164 and Rockwell-Collins AN/ARC-. The KTR-908 and KTR-909 radio stations from AlliedSignal (now Honeywell) subsidiary BendixKing are installed in the F-7MG prototype.

As the MiG-21 versions available to the Chinese (essentially the MiG-21F-13 type aircraft from the Soviet delivery and the MiG-21MF procured from Egypt) do not have the RSBN navigation system and Pion-N instrument landing system (these would only be integrated into the MiG-21bis), the corresponding equipment for the F-7 needed to be procured by importing Western technology. For the F-7MG this came from BedixKing and includes the navigation system KNR-634A, the TACAN KTU-709, the radio compass KDF-806 and the associated digital displays.

The US radio altimeter 0101-HRA/2 used from the J-7IIA/J-7IIM to F-7GS measures altitudes between 0 and 1,500 metres.

Early export versions have either the YD-3 (F-7A) IFF specially developed for export, the Type 602 device also used in the Chinese aircraft (F-7B, F-7IIK/F-7BK) or no such device (F-7IIN). The Pakistani versions of the F-7 are equipped with US AN/APX-101 IFF. The F-7GS is equipped with a Chinese Type 605A IFF.

In newer export versions, such as the F-7MB, F-7MP and FT-7P, the LJ-2 omni-directional RWR is installed to improve the self-protection capability. The F-7PG is equipped with the Chinese RWR ARW9101, which is also used in the J-10. The ARW9101 operates in frequency range from 0.7 Hz to 40 kHz, which covers the operating range of almost all existing radars. Signals from up to 100 radar models can be stored in the device's database. If the aircraft is detected by the enemy's radars, the system emits optical and acoustic warning signals and triggers the release of infrared and radar jammers.

The Soviet AP-155 dual-channel autopilot was probably the starting point for the development of the KJ-11 autopilot of the J-7III, which was tested from 1984 and later also used in the fourth-generation J/F-7.

With the procurement of Su-27s from Russia and its subsequent licence production as J-11 from 1997, China also gained access to the Sura-K helmet-mounted sight used in these aircraft, which is an export version of the Щель-3УМ (different spellings 3УМ and 3УМ) already produced in the Soviet Union for use in Su-27s and MiG-29s. Developed by the Kiev ZKB 'Arsenal' (now ARSENAL Special Device Production State Enterprise), the device essentially consists of the helmet-mounted sighting and target designation device and a position detection block with optical sensors that determine the pilot's line of sight. In the Su-27 and MiG-29, this information is transmitted both to the optoelectronic targeting system and to the missile seeker heads, thus speeding up target designation and shortening the time it takes to effectively engage the target. With the Щель-3УМ, target

A striking feature of the newer single-seaters is the angle-of-attack sensor, which has no moving parts. On the two-seaters, a model with a movable wing like on the MiG-21 is still used. (Holger Müller)

While the spherical antennas of the LJ-2 RWR are already part of the external appearance from the F-7M onwards, the flat VOR antennas only appear as standard on the J/F-7G. In production, these antennas are then mounted further up and to the rear of the vertical stabiliser. (Holger Müller)

In single-seaters, the countermeasure dispenser is integrated into the false keel as seen here; in double-seaters, this device is located between the two stabilising fins at the bottom of the rear fuselage. (Holger Müller)

assignment is possible between ± 60 degrees horizontally and from -15 to +60 degrees vertically; with the Sura-K, these values increase to ± 70 degrees and -35 to +65 degrees respectively. The helmet component weighs 390 grams, the other components between six and 10 kilograms.

A Chinese copy of the Sura-K is installed in the J/F-7 versions, which have a one-piece windscreen and thus the necessary visibility (J-7G, F-7PG/GS/BG/BG-I). At about the same time as a helmet-mounted sight becomes available in the J/F-7, this component is also being introduced in the MiG-21, namely in the Indian modernisation programme MiG-21UPG/Bison. This version receives the same Sura-K and is also equipped with a one-piece windscreen.

The one-piece windscreen of the J/F-7G generation finally allows the use of a helmet-mounted sight, which the struts of the classic MiG-21 cockpit had previously prevented. The helmet-mounted sight is a copy of the Sura-K, which China gained access to with the import of the Su-27. (CCTV)

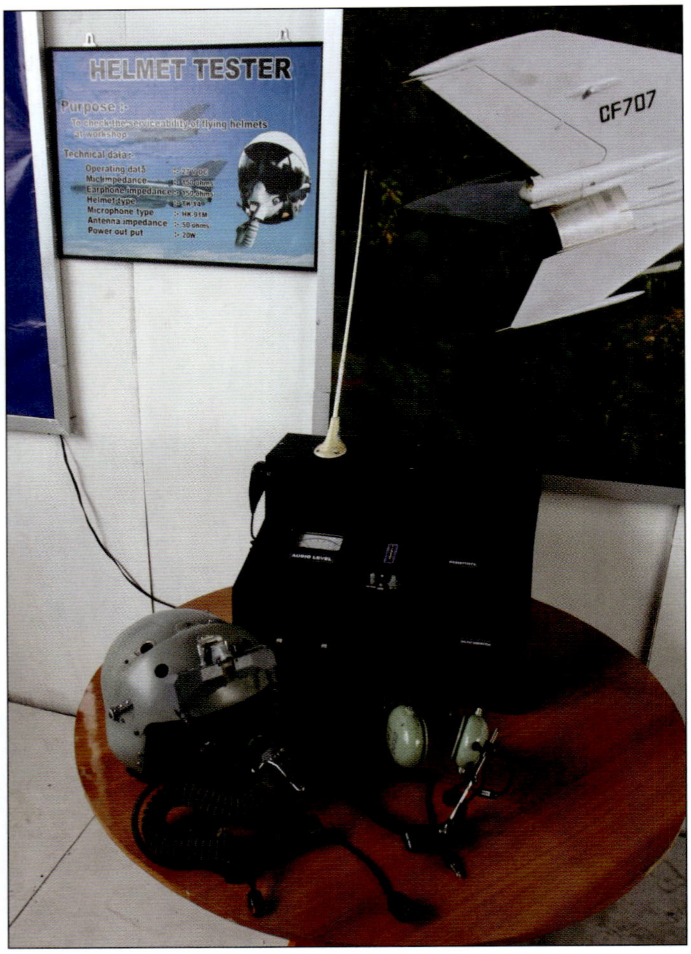

The helmet-mounted sight is then delivered with the J-7G-based export models, e.g. to Sri Lanka, where it is shown here together with the helmet and associated test equipment at the SLAF's 60th anniversary exhibition. (Holger Müller)

12 EJECTION SEATS

Ejection seats are used to accommodate the pilot in the cabin during the flight and to exit the aircraft in emergency situations.

SK

The SK ejection seat (*Sistema Katapultirowania*) was developed at the OKB MiG and used for the first time in the E-50/2 prototype. It allows ejecting at altitudes between 110 metres and service ceiling at speeds of up to 1,100 kilometres per hour. The specific parameters differ considerably in some cases, depending on whether the ejection process takes place protected by the canopy or without it. In the single-seat MiG-21 versions F, F-13, PF and FL, which have a forward-opening canopy with an integrated windscreen, this protects the pilot from the airflow during ejecting. The canopy is caught by the moving seat and remains connected to it until the airflow velocity falls below a critical level. The MiG-21U two-seaters, which are equipped with the SK seat too, have side-opening canopies that do not offer this option. Depending on the version, the maximum ejection speed without the protective canopy is up to 850 kilometres per hour, the minimum altitude without the canopy is 110 metres (at speeds above 500 kilometres per hour) or 130 metres (below 500 kilometres per hour) and 150 metres with the canopy. The integration of the canopy into the ejection process therefore increases protection for the pilot, but slows down the process.

The pilot begins the ejection process by tucking up his legs and then pulling the central ejection handle or the ejection handles on both sides of the seat (these were deleted during further development because of problems releasing the pilot from the seat). The leg restraints engage and the belts are tightened. The telescopic rods with the stabilisation parachutes extend. At lower speeds and altitudes, the canopy can be jettisoned, which shortens the process. At higher speeds and altitudes, the canopy remains in position until the seat moves up the guide rails in the cockpit, engages pins in the rear canopy latches, and takes the canopy with it. The canopy folds forward to an angle of up to 70 degrees, then it separates from the airframe and the forward latches also connect to the seat, so that it is positioned in front of the pilot, protecting his body. A pyro cartridge ignites and blasts the seat away from the aircraft with a maximum g-load of 18. The stabilised seat

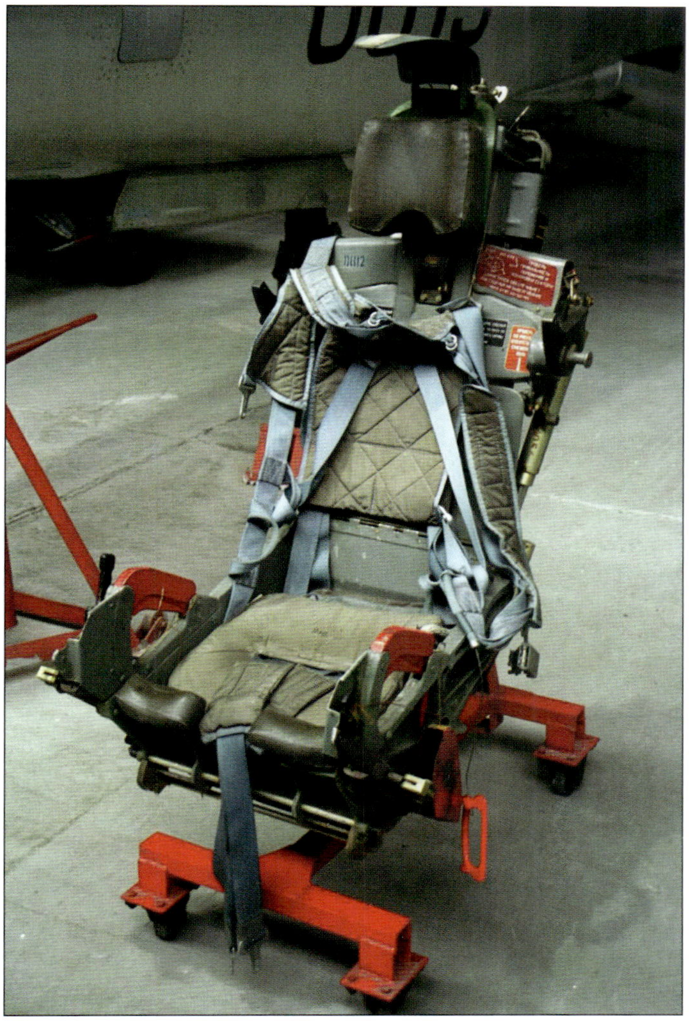

The interaction between the SK ejection seat of the MiG-21F-13 and the canopy make the escape system slow and error prone, which is why China was working on further developments at an early stage. (Holger Müller)

then drops until, on the basis of altitude or time after ejection, first the stabilising chute and then the cockpit canopy separate from the seat. Harness buckles and leg restraints open, and the pilot separates from the seat and descends to the ground by parachute.

HTY-2

J-7 and J-7I use the original ejection system SK of the MiG-21F-13, but the interaction of ejection seat and canopy is very complicated and comparatively slow, which affects the success rate, especially at low altitudes. From the Chinese point of view, the main problem with the SK – for which no conclusive explanation has yet been found – is the small distance between the pilot's head and the canopy.

As part of the development of the J-7I 'six modifications', an attempt was made to solve the problem of headroom by raising the rear of the cabin roof by 70 mm. The attempt is obviously not successful, as it is only implemented on a few aircraft. The author only has photo evidence of two aircraft.

A new – and ultimately successful – approach was pursued during the development of the J-7II, which began in 1975 and featured a rear-opening canopy with a fixed windshield. The canopy roof is raised by 50mm and the thickness of the parachute package on which the pilot sits is reduced, thus increasing headroom.

When ejecting, the canopy is always jettisoned first before the seat is ejected. The new Type II ejection seat (later designated HTY-2) uses a rocket engine with three-stage ignition instead of the previous pyro cartridge, which reduces the load on the pilot during ejection from 20g to 17g, increases the ejection altitude and optimises the flight path after leaving the cabin. A new two-stage stabilised deceleration chute was developed to slow down the seat stably at different flight speeds and to ensure that the pilot can be separated from the seat in time by the separation system. In addition, a central release handle was introduced – as in the later models of the SK. When it is actuated, the canopy ejection, seat ejection and pilot separation take place automatically.

Pulling on the central handle also ensures an optimum arm position, which prevents injuries caused by the airflow. At the end of the ejection process, the central pull ring is automatically released from the seat. This ensures that the pilot can detach from the seat without having to let go of the handle, which is impossible if he loses consciousness due to the high g-load. The minimum opening height of the new parachute is reduced from 100 metres to 60 metres. The parachute opening system has double redundancy. In addition, a new rucksack-like container is being developed that contains various equipment for both sea rescue and survival on land. The improved rescue system enables ejecting from zero altitude to service ceiling in the speed range between 250 and 850 kilometres per hour.

The same boundary parameters also apply to the use of the seat in JJ/FT-7 trainers. In order to prevent possible dangers due to overlapping flight paths and to protect the pilots from hot gases of the other seat, the two seats are ejected one after the other. The automatic ejection sequence runs as follows: drop of rear canopy, ejection of the rear seat, 0.4 second wait, drop of front canopy and ejection of the front seat. If the ejection sequence fails, the ejection process can be triggered manually.

The development of the modified rescue system for the J-7, consisting of the seat and the rear-opening canopy, was officially

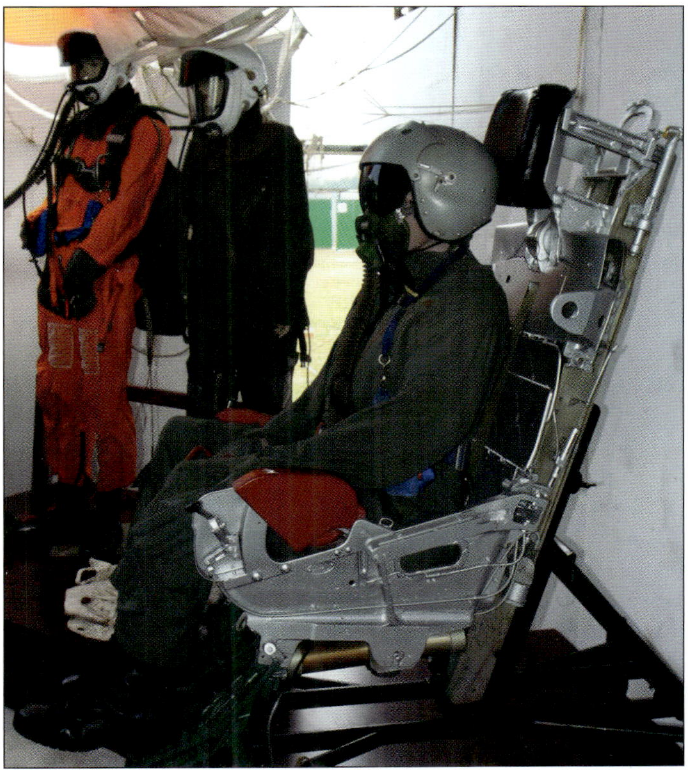

Despite its obvious shortcomings, the Chinese HTY-2 is based on the Soviet SK, from which the mechanical design was adopted largely unchanged. The main change is the separation of canopy and seat ejection. Further modifications include the parachute package, rocket propulsion and the additional central ejection handle. The latter modification was also introduced on the SK. (Holger Müller)

completed on 31 August 1979. The seat is said to have been used successfully abroad for the first time in 1984.

The HTY-2 is used in all single-seat indigenous versions from J-7II to J-7E as well as in the early production JJ-7 and JJ-7A two-seaters. Of the export variants, the first and second generation single-seaters and all 'short' two-seaters are equipped with it. In addition, 'long' two-seaters for less solvent customers, such as Iran, are also delivered with it.

HTY-4

Despite numerous improvements, the HTY-2 still has several shortcomings due to the largely unchanged design of the SK. For example, it does not have a zero-zero capability. As the parachute is located under the seat, the parachute ropes are only tensioned by the free fall of the pilot's body when the parachute is deployed, which means that the height loss of the opening canopy is relatively high.

New seat types were developed with the aim of enabling ejection at zero altitude even at low speeds. Between 1973 and 1977, Type III (later HTY-3) is developed for use in the J-8, which allows ejecting from zero altitude and in the speed range of 130 to 850 kilometres per hour. This is achieved by relocating the parachute package to the headrest, an auxiliary parachute deployed by pyro cartridge, which stabilises the seat and thus shortens the time until the main parachute opens, and a more automated ejection and separation system. In 1980, based on the Type III, the development of the Type IV for the J-7III begins. It differs from the original model in having a rocket engine with lower impulse, a modified auxiliary and main parachute and an increased seat height. The seat is available when flight testing of the J-7III begins in 1984. The minimum speed for ejecting is 140 kilometres per hour.

Both HTY-3 and HTY-4 still show strong similarities to the HTY-2 in terms of the structure of the seat frame, which certainly limits their performance.

J-7III and J-7IIIA are equipped with this ejection seat as standard. It is said to have been installed in the F-7MP as the only export version, but all available photographs show the HTY-2. In any case the Chinese seat in this version is replaced by the British Martin-Baker Mk.10L.

Martin-Baker Mk.10L

Pakistan chose the British Martin-Baker Mk.10L for its F-7 (as it had already done for the F-6), which had significantly better parameters than the Chinese seats of the late 1980s. The Mk.10L is a zero-zero ejection seat that allows ejecting up to speeds of 630 KIAS (approximately 1,166 kilometres per hour) and altitudes above 50,000 feet (15,250 metres). The seat consists of four main components: force-absorbing structure, seat shell, launch mechanism and parachute package.

When the release handle is pressed, the detonation cords of the canopy are triggered first (if present) or the canopy is blown off. Then the ejection cartridge, a telescopic tube with a two-stage charge at the rear of the seat, fires and the seat moves upwards in its guide rails. This activates the emergency oxygen supply and simultaneously disconnects all supply and communication links with the aircraft. The arm and leg restraints are activated.

As the seat continues to move upwards and out of the aircraft, the solid-fuel rocket engine ignites. A barostatic and a time-delay mechanism activate the main parachute below an altitude of 16,000 feet (4,876 metres), whereby the opening only takes place at sufficiently low speed. The seat is automatically separated from the pilot. Manual separation is possible if the automatic system fails.

All single (F-7P/PG) and double-seater (FT-7P/PG) versions of the PAF are equipped with the Mk.10L, which is designated PK10LZ in the Pakistani version. Bangladesh has procured the Mk.10L as BD10LZ for retrofitting its F-7MB and FT-7BB.

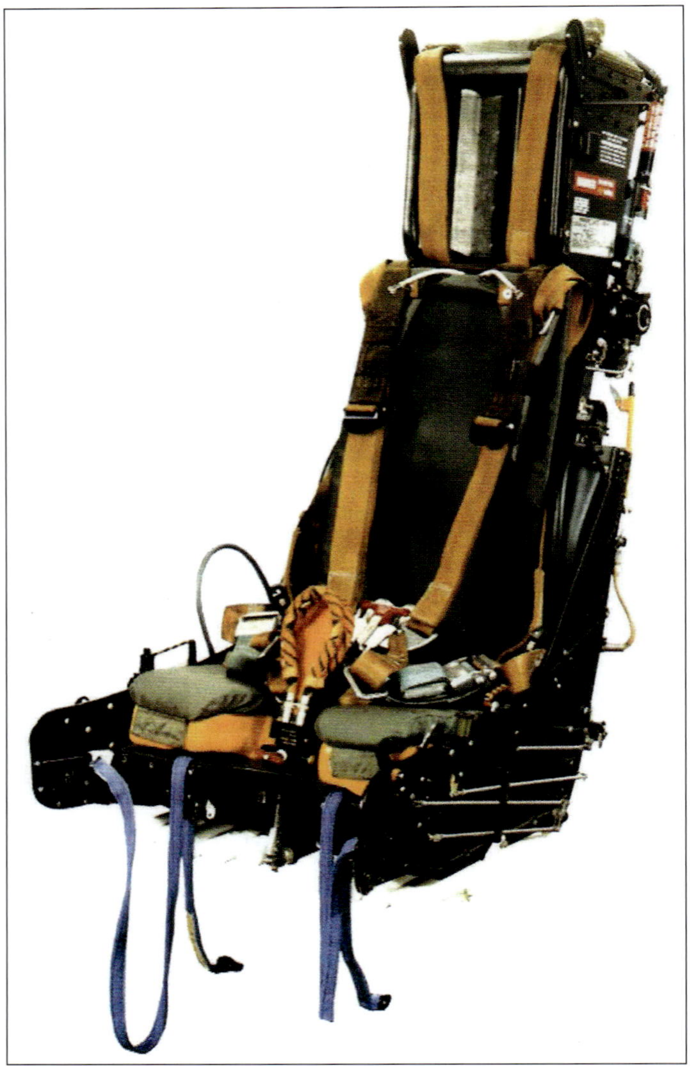

Although the Chinese worked intensively on the further development of the J/F-7 emergency escape system, their solutions had not reached the state-of-the-art in the 1990s. Pakistan, which already had Martin-Baker seats in service, therefore decided in favour of the Mk.10L in a version specially adapted to the F-7 under the designation PK10LZ. (Martin-Baker)

HTY-6/HTY-6M

The HTY-6 ejection seat (also known as TY6) developed by the China Aviation Life-Saving Equipment Research Institute (610) bears so many external similarities to the Martin-Baker Mk.10L that it can be assumed that the latter is at least the basis of the development. Thanks to Pakistan's procurement of Martin-Baker seats for its F-7, China has access to the corresponding technology.

The available data, such as the operational range at speeds of 0 to 1,100 kilometres per hour and altitudes from 0 metres to the service ceiling, also point to common genes.

The HTY-6 and its further development HTY-6M are installed in all fourth-generation J/F-7 single and twin-seaters with the exception of the aircraft destined for Pakistan. They are also used for retrofitting the J-7II, among others. The TY-6C variant for single-seaters with rear-opening canopies has short, flat canopy breakers, while the TY-6E for two-seaters with side-opening canopies has longer and more pointed breakers, with differences between the front and rear seats in terms of length and tip.

The HTY-6 ejection seat installed in the last F/FT-7 versions is so similar to the Mk.10L that it is not unreasonable to speak of it as a copy. The long and pointed canopy breakers on the HTY-6E two-seater version are striking. (Holger Müller)

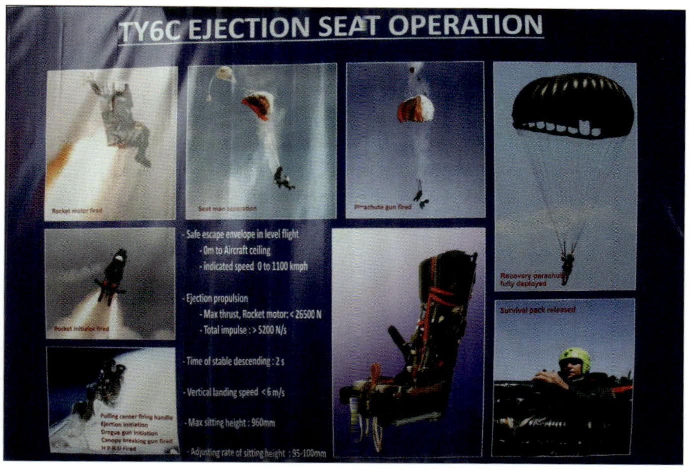

Illustration of functionality and performance data of the HTY-6 alias TY6. Here, too, the similarities to the Mk.10L are striking. (Holger Müller)

13 ARMAMENT

Cannons

The cannon armament of the J/F-7 consists of copies of Soviet models.

Type 30-1

The Model 30-1 is a copy of the Soviet single-barrelled NR-30 autocannon of the MiG-21F-13 by the Xi'an Kunlun Machinery Plant (Plant 847, today Xi'an Kunlun Industrial Company). The 30mm calibre gun consists of six main components and a total of 163 individual parts. It is simple in design and easy to dismantle. It is 2.15 metres long, 0.18 metres wide, 0.186 metres high and weighs 66 kilograms. The initial velocity of the bullet is 780 metres per second and the rate of fire is 850 to 1,000 rounds per minute. The barrel has a service life of 2,000 rounds.

When the gun is fired, the energy of the recoil is used to open and return the bolt and to eject the case. The energy of the powder gases, which are channelled out of the barrel into a gas collector during firing and compressed during the backward movement, is used for braking (there is also a hydraulic buffer and a spring mechanism for this purpose) and forward movement of the barrel as well as for feeding the cartridge. The powder gases are then released into the outside air. Compressed air is used for cocking and firing is triggered electrically.

The ammunition is fed by belt. Seven different types of ammunition are available for use against air and ground targets. These include fragmentation incendiary (OFZ) and armour piercing types (BR), ammunition for combating balloons and a radio interference projectile (PRL) filled with chaff.

Due to the strong recoil of the gun, the gun vibrates considerably during continuous firing, which affects the alignment between the barrel and the ammunition feed and thus reduces the reliability. As a result of the barrel recoil principle, moving parts make up a large part of the weapon's mass, which increases wear. The design of the gas cylinder makes it difficult to remove combustion residues and thus complicates maintenance. The ammunition supply is 60 rounds per gun.

All single-seat variants of the J/F-7 with the narrow air intake – with the exception of the J-7EB/GB demonstration versions – are equipped with one or two 30-1 cannons.

Reloading ammunition into the left cannon on a Chinese J-7. While the mechanic, who is only visible in the background, pulls the belt into the fuselage with a winch, his colleague under the aircraft ensures that the projectiles do not hit the airframe. (Xi Bobo)

A Chinese J-7G with a censored serial shows one of its many possible load variants. The aircraft carries a 720-litre auxiliary fuel tank under the fuselage, and HF-7 series pods for unguided 90mm 90-1 rockets on the inner wing pylons. The outer wing hardpoints, which can accommodate both weapons and auxiliary fuel tanks, are unused. It is unclear whether – as with the MiG-21 – these pylons would have to be exchanged to change the load. The countermeasure launchers under the tail are partially loaded. As can be seen from the lack of smoke marks, neither cannon nor missiles have been fired so far. (Yang Pan)

Although this version has only one gun, two 30-1 cannons are displayed under this F-7GS. In comparison with the AAMs next to it, the considerable length of the weapon of 2.15 metres becomes clear. (Holger Müller)

A J-7II engages ground targets with the left 30mm cannon. The considerable muzzle flash, the ejected shell casings (under and behind the left wing) and the trail of powder gases trailing behind the aircraft are clearly visible. (Xi Bobo)

Type 23-3

When the GSh-23 cannon of the MiG-21MF became available, it was also copied by the Xi'an Kunlun machine factory under the designation 23-3 (the suffix -3 is the consecutive number within the type designation – the Chinese versions of the Soviet NR-23 and AM-23 are designated 23-1 and 23-2 respectively).

The 23-3 is a double-barrelled 23mm calibre automatic cannon with a rate of fire of 3,000 to 3,400 rounds per minute, which is significantly higher than that of single-barrelled cannons of the same calibre (the single-barrelled 23-1 cannon only achieves 850 rounds per minute). Despite the higher rate of fire, the volume and weight (50.5 kilograms without muzzle brake) have not increased significantly. The achievable rate of fire is close to that of a revolver cannon of the same calibre with a much simpler installation in the aircraft.

Both barrels of the gun as well as the entire automatic system for feeding the cartridge, locking, firing and ejecting the case are located in a common housing, whereby firing takes place alternately. The mechanics are forcibly interconnected. The gun utilises the energy of the powder gases to move the mechanism, whereby the gases are first extracted from one barrel and then from the other. The shells are fed from a belt via a sprocket-shaped mechanism. The cannon has an electrically operated fire control and cocking mechanism, with the latter triggering a cocking cartridge.

The cover of the Type 23-3 cannon is integrated into the central speed brake of the FT-7. Unlike the original GSh-23 on the MiG-21, the Type 23-3 is equipped with a muzzle brake to reduce recoil. (Holger Müller)

The gun fires 23×115mm shells with fragmentation projectiles, armour piercing projectiles, fragmentation incendiary tracer projectiles and radio interference projectiles with DOS-4 or DOS-15 chaff as well as incendiary projectiles with highly sensitive fuses for destroying balloons. The usual belting consists of fragmentation and armour piercing projectiles in a 1:1 ratio.

When engaging aerial targets, aiming is carried out using the radio or optical sight; when engaging ground targets, the optical sight is used exclusively.

In the single-seater versions J-7III and J-7IIIA as well as in the stretched two-seaters from FT-7P onwards, the cannon is permanently installed. In this case, the ammunition supply, which is housed in bays under the fuselage skin, contains 200 rounds. JJ-7A and JL-9 can carry the gun in a container under the fuselage.

Guided Air-to-Air Missiles

Due to its radar and avionics equipment, its payload and the technology available in China at the time of its development, the J/F-7 can only use short-range infrared missiles. It is not equipped with medium- or long-range radar-guided missiles.

PL-2

As the first Chinese air-to-air missile PL-1 (pronounced *PiLi*, thunder) was only produced in small series and only made available for the J-6 (MiG-19), the first missile armament of the J-7 was the infrared-guided PL-2 air-to-air missile, a copy of the Soviet K-13 (R-3S), which in turn was a copy of the early AIM-9B Sidewinder. During the air battles over the Taiwan Strait in 1958, Taiwanese F-86s used this then state-of-the-art air-to-air missile for the first time. At least one undamaged copy of the missile ended up in the hands of the Chinese and was eventually passed on to the Soviet Union, where it was copied due to the lack of comparable indigenous weapons.

Production documents and pattern parts of the R-3S were handed over to China in 1961 together with the deliveries for the licence production of the MiG-21. Work on licence production began in 1964 and in November 1967 the missile, now known as the PiLi-2, was transferred to series production at the Xi'an Dongfang machine factory (Plant 844, now Xi'an Dongfang Electromechanical Group Company). As with the J-7, however, it took several years before any significant output was achieved.

For the two-seaters with the short fuselage, the Type 23-3 cannon is available in a pod mounted on the fuselage hardpoint (similar to the 12.7mm machine gun pod of the Soviet trainers). It is not known whether this was also exported. (dlsunshy)

Two examples of the PL-2 in the Beijing Military Museum. The seeker of the lower missile is covered by the protective cap, which is only removed before the aircraft takes off. At first glance, the PL-2 can only be distinguished from the Soviet R-3S by the different labelling. (Chinese Internet)

The PL-2 has a canard design with cross-shaped control surfaces and trapezoidal tail fins. The missile consists of the following components: infrared seeker head, warhead with optical-proximity fuse, control unit for vertical and horizontal flight control with rudder actuators, energy supply unit, and rocket motor. The PL-2 has a length of 2.84 metres, a diameter of 0.127 metres, a wingspan of 0.609 metres and a weight of 75.3 kilograms.

The missile can only be used in simple weather conditions from the rear hemisphere. The infrared seeker head based on a lead sulphide detector has a detection angle of 3.5 degrees. The missile must have acquired the target before launch, which requires precise alignment of the carrier aircraft with the target due to the narrow detection angle. After acquisition, the target can be tracked within a 25-degree cone, but launch is only possible with not exceeding 2g and at speeds greater than 800 kilometres per hour. The maximum firing distance is between 1.5 and 3.5 kilometres, depending on altitude, and the range is between 4 and 7.6 kilometres. When firing in the direction of the sun, a minimum angle of 20 degrees to the sun must be maintained. The delta-shaped, fully manoeuvrable control surfaces with swept trailing edges each lie in the same plane as the respective tail fin. Impact fuses are located on their leading edges to trigger the warhead in the event of a direct hit. The control motors are powered by powder gases tapped from the rocket motor.

The rocket has a blast fragmentation warhead. The outer skin forms the warhead's fragmentation liner, so that it is produces a directional fragmentation pattern. Behind the warhead is the optical-proximity fuse with a range of nine metres, which reacts to the target's thermal signature.

The solid-fuel engine with an axially symmetrical thrust nozzle is located at the rear of the rocket. Four trapezoidal tail units with straight trailing edges, each at an angle of 45 degrees to the horizontal plane, are arranged at the end of the fuselage.

The solid-fuel propulsion system only operates for two seconds and accelerates the missile to a speed of around 700 metres per second. The agility of the missile is very low, so that in practice only non-manoeuvring targets can be engaged. The warhead, which weighs 11.4 kilograms and generates approximately 1,000 fragments, has an effective radius of 10 to 11 metres, but the sensitivity and resistance to countermeasures of the infrared proximity fuse are low. If the missile misses the target, it self-destructs after 25 to 26 seconds. The controlled flight lasts a maximum of 21 seconds.

It is unclear whether an inert PL-2 missile similar to the Soviet R-3U is available for training purposes. The latter has a seeker head, but not a tail unit, warhead and engine. The inert missile delivers the same signals as the live projectile when it is targeted and also records data for subsequent evaluation.

In the late 1970s, a version with an improved seeker and increased resistance to countermeasures was produced under the designation PL-2A. With its conical front section, the PL-2A is externally similar to the American AIM-9D Sidewinder or the Soviet R-13M, but the seeker is considerably smaller and is not cooled. The control surfaces and tail unit remain unchanged compared to the PL-2, the fuselage is slightly longer.

The final version of the missile is the PL-2B with the control surfaces moved further forward and a larger warhead. Development began at the China Air-to-Air Missile Research Institute (Institute 612, now the Luoyang Opto-Electro Technology Development Centre) in October 1964 under the code name PiLi-2-15. The sensitivity of the infrared seeker was improved, the weight of the projectile and the delay time of the proximity fuse were reduced. The infrared sensor has a higher detection range and greater resistance to interference from the sun and sky background. The sensitivity of the optical detonator and its reliability are increased. The PL-2B was transferred to series production in October 1981 and forms the most important missile armament of the J-7II in the 1980s. Variants of the PL-2 can be used by all first and second generation J-7s as well as by the two-seaters.

PL-5

The development of the PL-5 at the Air-to-Air Missile Research Institute from the mid-1960s onwards was again based on a Sidewinder version, namely the AIM-9D, of which, however, only debris was available. Later, the AIM-9L – possibly from Pakistani stocks – was apparently also available as a basis for development. The external resemblance of both the missile and the launch rail clearly point to this origin. The lead sulphide seeker head is cooled by compressed air to achieve a higher sensitivity and has a better resistance to countermeasures than the seeker head of the PL-2.

The design of the PL-5B (the development of the original PL-5A was stopped after years of development) was completed in September 1986. Series production took place at the Hanzhong Nanfeng Machinery Factory (Plant 202).

Due to the barrier fence, it is only recognisable at second glance that the FT-7 exhibited at the Aerosalon in Paris-Le Bourget 1987 carries a PL-2A under the wing. Apart from the type-designation on the missile body, the conical front section identifies this version. (Guido E. Bühlmann)

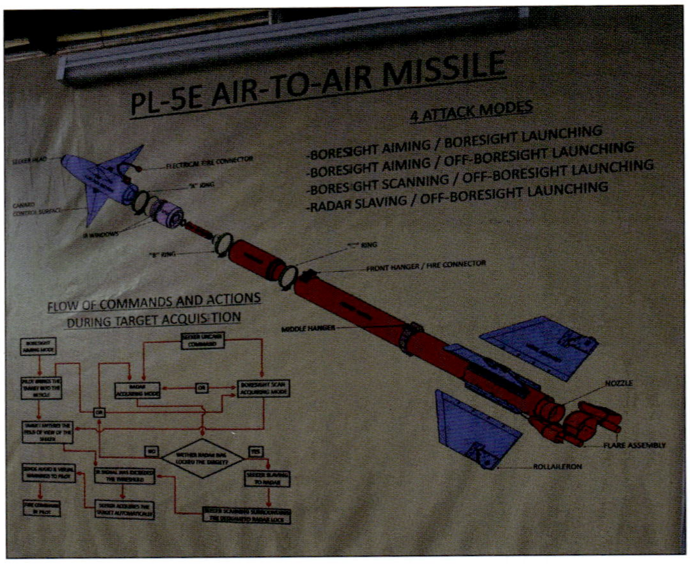

Schematic of structure and function of the PL-5E. (Holger Müller)

The aerodynamic layout consists of four components: the control section, the warhead, the fuse and the rocket motor. The double-delta control surfaces are located at the front of the projectile, the trapezoidal wing with gyro rudder at the rear, each arranged in a cross shape. The PL-5B has a length of 2.892 metres (according to other sources 3.128 metres, which seems unlikely in view of the available image material), a diameter of 0.127 metres, a wingspan of 0.657 metres and a weight of 87 kilograms. The maximum range is five kilometres, the minimum range 1.3 kilometres. The solid-fuel propulsion system gives the missile a maximum speed of Mach 2.2 and the maximum g-load is 30. Two types of warheads are used – fragmentation type and continuous rod type. The former is equipped with an infrared proximity fuse and the latter with a radio proximity fuse. Both types of warheads and their proximity fuses have a total mass of six kilograms and an effective

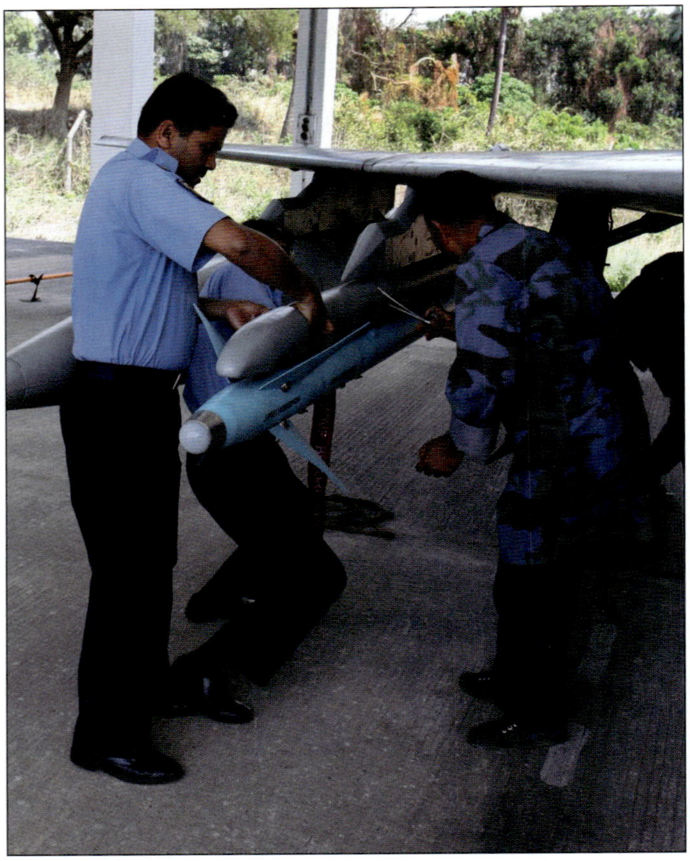

A PL-5E-II training missile mounted on an F-7BG-I. The launch rail, which looks very similar to the US LAU-7 for the Sidewinder, holds the nitrogen container for cooling the seeker head. (Holger Müller)

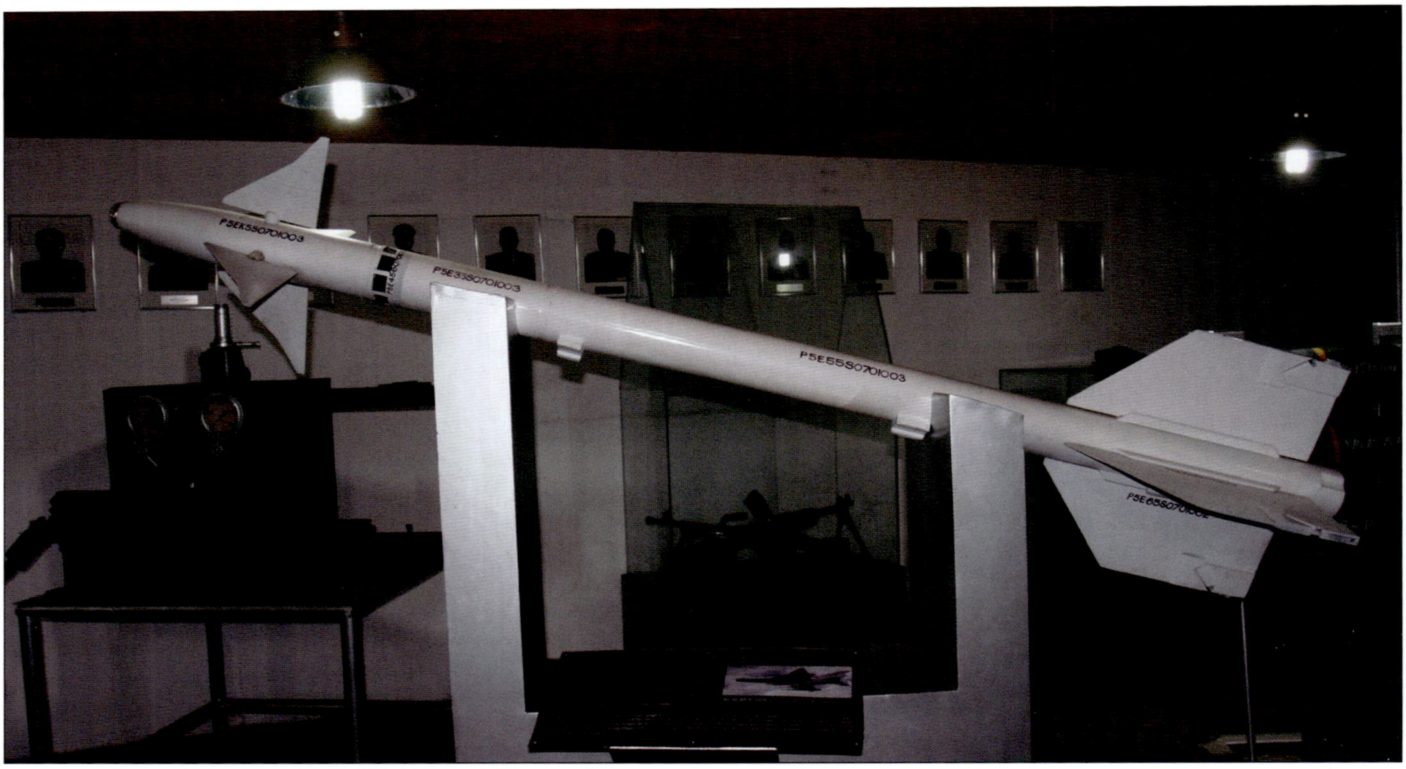

A PL-5 at the Sri Lanka Air Force Museum in Ratmalana. The similarity to the AIM-9L Sidewinder developed in the mid-1970s is obvious. (Holger Müller)

range of 10 metres. The infrared proximity fuse is highly resistant to countermeasures and has a high detonation reliability.

Utilising the technologies used in the PL-7 and PL-8 (see there), a much more powerful version, the PL-5E, was developed at the end of the 1980s as a low-cost alternative to the aforementioned missiles. The indium antimonide infrared seeker, cooled with liquid nitrogen, has a deflection of 25 degrees before and 40 degrees after launch. The missile has a smokeless rocket motor which reduces interference with the engine of the carrier aircraft. The fragmentation warhead with a total mass of nine (PL-5E) or 11.5 (PL-5E-II) kilograms is triggered by a laser proximity fuse. The projectile has a diameter of 0.127 metres, a length of 2.89 metres, a wingspan of 0.617 metres, a weight of 83 kilograms, a maximum speed of Mach 2.5 and a maximum manoeuvring g-load of 40. The range is now 16 kilometres, for the enhanced PL-5E-II 18 kilometres.

While the PL-5B can only be fired from the rear hemisphere, the PL-5E can be used from any direction. The PL-5B is compatible with domestic versions from J-7III onwards and export versions such as the F-7MP, although there is no evidence for export. The PL-5E is part of the armament of all J-7G-based export versions.

PL-7

The limited performance of the PL-2 and PL-5B seeker heads makes it almost impossible to use these missiles in manoeuvre air combat. This was remedied by access to Western missile technology, which became possible from the mid-1970s onwards thanks to China's improved relations with the West. The PL-7, officially developed by Zhuzhou Southern Aviation Power Machinery Co., Ltd. is a result of this cooperation. This infrared-guided air-to-air missile, first presented to the public in 1987, is identical to the French Matra 550 Magic. It is unclear whether this is a licensed design or a relabelling of the original.

Right: The PL-7 is used exclusively to arm the F-7 for export. Together with the live missiles, practice weapons and dummies for handling training – here an Iranian example – were supplied. (Holger Müller)

Below: Interestingly, the same Western missile models equip both the F-7 for export and the modernised MiG-21 variants, the Matra R.550 Magic alias PL-7 (front) and the Rafael Python 3 alias PL-8 (rear). Here both missiles could be seen together with a Soviet R-60 in front of a MiG-21 Lancer C. (Holger Müller)

The PL-7 has a high g-load capability and can engage targets at high off-boresight angles. The missile is 2.75 metres long, has a diameter of 0.157 metres, a wingspan of 0.66 metres and weighs 89 kilograms. The maximum range is 14 kilometres, the minimum 0.5 kilometres. The missile reaches a maximum speed of Mach 2 and a maximum g-load of 35. There are four fixed trapezoidal wings at the head of the canard-controlled missile, followed by four movable control surfaces of rectangular shape with an attached triangle at the end. The tail unit of the missile is attached to a sleeve that rotates around the longitudinal axis of the projectile in order to stabilise it. The control surfaces are arranged in a cross shape and are in the same plane. The directional control of the missile is achieved by deflecting the rudders in pairs, the roll control by differential deflection. The canard rudders are located in front of the missile's centre of gravity and interact with the large wing mounted on the tail, which enables high manoeuvrability and g-loads in close air combat. The projectile consists of two main assemblies. The front part contains the seeker, electronic components, the servomechanism, the battery, the control unit and the infrared proximity fuse. The rear section houses the warhead and solid propellant rocket motor. The infrared seeker uses a nitrogen-cooled indium antimonide infrared detector with high sensitivity and a maximum detection angle of ± 30 degrees. The warhead has a total weight of 12 kilograms and an infrared proximity fuse with an effective range of 10 metres. The PL-7, which went into series production in 1987, is intended exclusively for export variants of the J/F-7. It was not adopted by the PLAAF.

PL-8

Another originally foreign product is the PL-8, the most powerful Chinese air-to-air missile at the time of its introduction at the end of the 1980s, a licensed version of the Israeli Rafael Python 3. The highly agile passive infrared-guided missile is used to engage air targets of all kinds from all directions, against any background and during active manoeuvres by the target. The warhead is triggered by an active proximity fuse.

The missile has a length of 2.99 metres, a diameter of 0.16 metres, a wingspan of 0.86 metres and a weight of 120 kilograms. The maximum range is 15 kilometres, the minimum 0.5 kilometres, the maximum speed Mach 3 and the maximum g-load is 40.

The missile is canard-controlled, with aerodynamic rudders at the front and four large wings at the rear. The fuselage is divided into two parts. The front section contains the seeker head, electronic components, batteries, nitrogen tanks and the active radar proximity fuse, while in the rear section warhead, control motor and rocket engine are located. The seeker uses a nitrogen-cooled indium antimonide infrared detector and detects targets at an angle of ±20 degrees off the boresight of the launching aircraft. The warhead weighs 11 kilograms and has an effective radius of 13 metres. The large swept wings are not mounted directly on the tail but slightly offset towards the centre. At the trailing edge of each wing is a roll-stabilising gyro rudder that rotates at up to 10,000 revolutions per minute under aerodynamic force. The use of a highly sensitive infrared detector seeker, active radar proximity fuses and a two-stage solid-fuel rocket motor significantly improves the missile's detection field, accuracy and range. The weapon's performance matches or exceeds that of the American AIM-9L Sidewinder.

Initially manufactured from imported CKD kits, the PL-8s are later produced entirely in China. Targeting can be carried out either by radar or helmet-mounted sight. The Python 3 supplied directly from Israel is known in China as the PL-8, while the PL-8A and PL-8B are missiles manufactured in China from Israeli parts kits or built entirely from Chinese components, respectively. The missile is manufactured by Zhuzhou Southern Aviation Power Machinery Co., Ltd. and equips the J-7H, J-7E and J-7G. It is not exported.

PL-9

The PL-9 is developed at the same time as the PL-8. Export restrictions for the PL-8 are thought to be the reason for the parallel development of two air-to-air missiles with identical purposes. The PL-9 combines the seeker technology of the PL-8/Python 3 with a Chinese missile body.

Developed by the Luoyang Solar Electric Technology Development Centre and the Xi'an Dongfang Electromechanical Group Co., Ltd., the PL-9 has similar dimensions as the PL-8, but resembles the PL-5B in aerodynamic design. At the front of the missile are small double-delta canards, while large trapezoidal wings with gyro stabilisers are fitted at the rear. As with the PL-7 and PL-8, the missile body is divided into two compartments to facilitate maintenance. The missile has a length of 2.99 metres, a diameter of 0.16 metres, a wingspan of 0.810 metres, a weight of 123 kilograms, a maximum range of 16 kilometres, a minimum range of 0.5 kilometres, a maximum speed of Mach 3.5 and a maximum g-load of 40. The missile is equipped with an infrared seeker cooled by liquid nitrogen. The PL-9 has automatic search and intercept functions and withstands high g-loads when manoeuvring. It can be used against targets off boresight. The warhead weighs 12 kilograms and is equipped with an active radar proximity fuse.

The PL-9 can be used in combination with a helmet-mounted sight. The missile has been in series production since 1989 and is available for indigenous and export versions of the J-7G.

AIM-9L/AIM-9P Sidewinder

The US-made Sidewinder versions that equip Pakistan's F-7s are based on the first guided air-to-air missile ever to be built in large numbers. Its first version entered service in 1956, since when the missile has been continuously developed and is still in production today in a completely revised form.

Sidewinder versions used by the F-7 (and MiG-21) include the AIM-9L and AIM-9P3. The AIM-9L is a much improved version of the Sidewinder, which went into series production in 1977. For the first time, a cooled seeker allows targets to be detected from the front hemisphere. The control surfaces have a double delta shape. The warhead is triggered by an active laser proximity fuse.

The AIM-9 has a canard design with cross-shaped control surfaces and trapezoidal tail fins. The missile consists of the following components: indium antimonide infrared seeker head, warhead with laser proximity fuse, control section for vertical and horizontal flight control with rudder drive, power supply unit and engine. The AIM-9L has a length of 2.89 metres, a diameter of 0.127 metres, a wingspan of 0.64 metres and a weight of 84 kilograms. The maximum firing range is six kilometres from the rear hemisphere and three kilometres from all other directions. The maximum range is 18 kilometres, the maximum speed Mach 2.5 and the maximum g-load is 30.

The AIM-9P3 was developed as a reduced-power export variant to complement the AIM-9L. It has a low-smoke engine and the same proximity fuse as the AIM-9L, but with a less sensitive seeker head.

Unguided Rockets

In addition to guided missiles and fuel tanks, two pods for unguided rockets (but obviously not four, unlike the MiG-21) can be mounted on the J-7's underwing pylons.

Pakistan has equipped its first-generation F-7s exclusively with US armament such as the AIM-9P Sidewinder, not least because the corresponding Chinese weapons systems were still far from Western standards in the early 1990s. Next to the Sidewinder are CBU-100 anti-tank cluster bombs (white), Mk.82 free-fall bombs (black), practice bombs (light blue) and belts with 23mm shells for the Type 23-3 cannon. (Holger Müller)

Different pods for unguided 57 mm 57-2 rockets on and below a F-7GS. Mounted under the wing is the HF-5A for 18 rockets, in front of the aircraft are two HF-18S for 12 rockets each. (Holger Müller)

57-2

The 57-2 57mm calibre unguided rockets are based on the Soviet S-5 and designed to engage single and area targets on the ground as well as air targets. They consist of a steel body with a warhead and impact fuse at the tip and a solid propellant rocket engine with an elongated thrust nozzle and the forward-folded stabilising surfaces with a span of 0.23 metres attached to it. These unfold when leaving the launch tube. While the Soviet models have eight of these surfaces, only six of them are found on Chinese rockets. Their maximum range of five kilometres is also said to be significantly greater than the two kilometres of the Soviet S-5.

The maximum speed of the 57-2 rocket is 510 to 610 metres per second and its length is 0.805 to 0.818 metres. The projectiles weigh between 3.86 and 3.97 kilograms, of which the warhead accounts for 1.38 kilograms. There are explosive, cumulative (shaped charge), armoured incendiary and incendiary warheads with impact or proximity fuses. The rockets are fired from various pods. The HF-5A, which outwardly resembles the Soviet UB-16-57UDM with its protruding launch tubes, can hold 18 rockets. The HF-18S which resembles the original Soviet ORO-57KM with a straight front, and the HF-16A each hold 12 rockets, of which either salvos of four rockets or all 12 can be fired.

90-1

The 90-mm 90-1 rocket, which is used exclusively to engage ground targets, is a indigenous Chinese development without a Soviet equivalent. Its structure is similar to that of the 57-2 rocket, but the

PLAAF ground crew load a pod with 57-2 rockets. The technician in the foreground pushes the missile into the launch tube, his colleague behind him connects the firing contacts. (Xi Bobo)

Air-to-ground firing of a Chinese J-7II with unguided 57mm rockets. While the stabilising surfaces of the projectiles that have just left the launch tubes are still folded, they have completely unfolded on the rockets in front of the aircraft. (Xi Bobo)

The HF-7 rocket pod holds seven of the much larger 90-1 rockets. There was no direct Soviet predecessor for this model. The similar S-8 with a calibre of 80mm was only developed after the Sino-Soviet split. (Holger Müller)

A J-7G of the PLAAF firing 90-1 rockets. The smaller number of stabilisation wings (four according to Chinese specifications, although only three surfaces are visible here), which, unlike on the 57mm rockets, only fold out to around 45 degrees, can be clearly seen. (Yang Junpeng and Yang Pan)

stabiliser at the rear with a wingspan of 0.34 metres has only four folding wings. The solid-fuel rocket motor gives the projectiles a maximum range of seven kilometres and a maximum speed of 500 to 600 metres per second. The length of the rocket is 1.257 to 1.267 metres and the weight is 14.6 to 17.1 kilograms, of which 4.8 to 5.6 kilograms are accounted for by the warhead with impact or proximity fuse. The 90-1 rocket is fired from pods of the HF-7 series. Each container can carry seven rockets, two or all of which can be fired at once.

Bombs

The bomb armament of the J/F-7 includes free-fall bombs, guided and cluster bombs of both Chinese and Western origin in weights of 50, 100, 250 and 500 kilograms.

The bombs are mainly high explosive and fragmentation bombs. Two bombs of 250kg or 500kg weight can be attached to the versions with two wing pylons, four bombs of 250kg or two bombs of 500kg to the versions with four wing pylons (500kg bombs at the inner pylons only). They are attached to universal pylons with bomb locks (one bomb per pylon).

The F-7BG-I (and thus probably also the J-7G) has the ability to drop bombs with the LS-6 (*LeiShi*, Thunderstone) satellite-controlled guidance kit. The kit consists of two parts: a fuselage with fold-out wings, which is placed on the back of a free-fall bomb, and a tail module with stabilising fins. This module contains an inertial navigation system, the flight controls and a battery. The satellite control system can be based on the US GPS, the Russian GLONASS or the Chinese BDS. If none of these systems is available, inertial navigation is used for control. The wingspan of the 500kg variant is 2,400mm and the maximum range is 60 kilometres from an altitude of 11,000 metres. The Bangladesh Air Force even states a range of 100 kilometres for the LS-6.

In principle, the J/F-7 should also be able to use the LT-2 laser-guided glide bomb, but only with target illumination on the ground or by another aircraft due to the lack of a laser target pod. The LT-2 is China's first mass-produced guided bomb, consisting of a 500-kilogram free-fall bomb and the guidance kit comprising the guidance head and tail unit. The bomb uses the same guidance mechanism as the American Paveway I/II and is probably based on it. The range of the LT-2 is between 10 and 15 kilometres, with a target deviation of less than 6.5 metres.

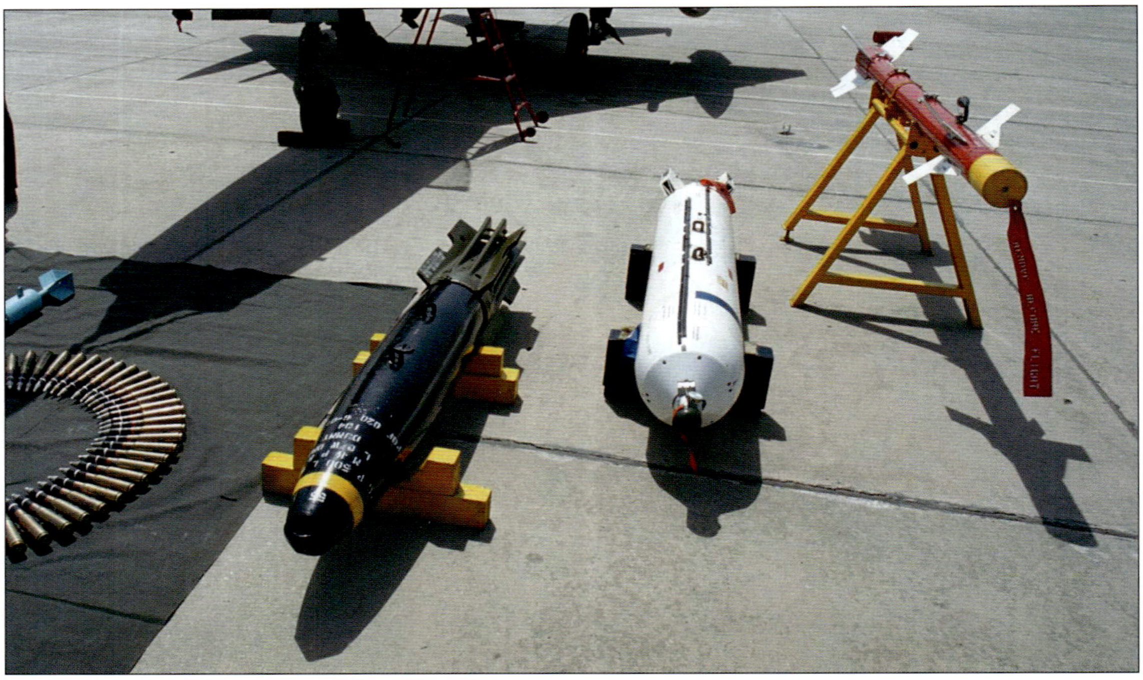

Top: Chinese Type 3 250kg low-drag, free-fall bombs with Pakistani AB 103 fuzes. (Holger Müller)

Centre: In addition to the normal Mk.82 free-fall bombs, Pakistan also has the Snake Eye variant with deployable air brakes (dark green tail section) in its inventory. Also on display here are 30mm shells for the 30-1 cannon, a CBU-100 cluster bomb and a Sidewinder practice missile. (Holger Müller)

Left: Guided bombs from Chinese production are available for the latest F-7 generation, such as the LS-6 satellite-guided bomb kit, a model of which is shown here at the CATIC stand at the NAF Air Expo 2010. (Holger Müller)

Left: Nigerian pilot with TK-11 helmet and anti-g pants made in China. The leg garters worn over the pants are to be attached to the HTY-6 ejection seat. (Holger Müller) Right: Survival suits made in China are also available for flights over the sea. (Holger Müller)

Above left: Components that are supplied with the aircraft, but are not part of the aircraft itself, include pilot equipment, ground equipment and accessories. The Chinese TK-1 high-altitude helmet, which was exported even with the latest F/FT-7s (right), is obviously based on the Soviet GSh-4 from the 1950s. The green cap (centre) is worn under this helmet. The standard TK-11 pilot's helmet (left), on the other hand, is more reminiscent of Western models and has a mount for the HMS. (Holger Müller)

Above right: Like the MiG-21, the J/F-7 also has two on-board batteries that enable autonomous engine start without outboard power supply. Externally, the Chinese battery models are largely identical to the Soviet ones. (Holger Müller)

Right: The Chinese brake chute solution differs not only in the design of the container, but also in the design of the chute package. In the Chinese version, it is shorter and the attachment eyelet is designed differently. (Holger Müller)

SELECTED BIBLIOGRAPHY

A 187 / 1 / 123 Flugzeug 66 - Zelle: Nationale Volksarmee der DDR 1968

A 187 / 1 / 128 Flugzeug MiG-21PF - Betriebsvorschrift - Teil I - Zelle - Triebwerk - Bildband: Nationale Volksarmee der DDR, Kommando der LSK /LV 1967

Anon, *China Today: Aviation Industry* (Beijing: China Aviation Industry Press, 1989)

Brent, Winston, *African Air Forces* (Nelspruit: Freeworld Publications, 1999)

Cooper, Tom and Bishop, Farzad, *Iran-Iraq War in the Air* (Atglen: Schiffer, 2000)

Cooper, Tom and Weinert, Peter, with Fabian Hinz and Mark Lepko, *African MiGs Volume 1 – Angola to Ivory Coast* (Houston: Harpia Publishing, 2010)

Cooper, Tom and Weinert, Peter, with Fabian Hinz and Mark Lepko, *African MiGs Volume 2 – Madagascar to Zimbabwe* (Houston: Harpia Publishing, 2011)

Cooper, Tom, Babak Taghvaee, and Liam Devlin, *IRIAF 2010* (Houston: Harpia Publishing, 2010)

Davies, Steve, *Red Eagles: America's Secret MiGs* (Oxford: Osprey Publishing, 2008)

Fontanellaz, Adrien, *Paradise Afire Volume 3: The Sri Lankan War 1990 – 1994* (Warwick: Helion & Company 2020)

Hewson, Robert, *Jane's Weapons: Air-Launched 2014 – 2015* (Coulsdon: IHS, 2014)

Kim, Kuk-chŏl (ed), *Immortal history of DPRK-PRC friendship* (Pyongyang: Foreign Languages Publishing House of the DPR Korea, 2022)

Matoulek, Ovčačik, Pavlovský, Susa, *MiG-21F/U* (Prague: 4+, 2008) (in Czech)

Mitzer, Stijn and Oliemans, Joost, *The Armed Forces of North Korea* (Warwick: Helion & Company, 2020)

Müller, Holger, *MiG-21* (2nd edition) (Stuttgart: Motorbuch Verlag, 2022) (in German)

Müller, Holger, *MiG-21 in action – Part 1: South and Southeast Asia: FliegerRevue Extra 21* (Berlin: Möller Buch und Zeitschriften Verlag, 2008)

Müller, Holger, *MiG-21 in action – Part 2: Middle East and Arabia: FliegerRevue Extra 23* (Berlin: Möller Buch und Zeitschriften Verlag, 2008)

Müller, Holger, *MiG-21 in action – Part 3: Black Africa and America: FliegerRevue Extra 25* (Berlin: Möller Buch und Zeitschriften Verlag, 2009)

Müller, Holger, *Sri Lanka's Air Force turned 60: FliegerRevue Extra* (Bergkirchen: PPV Medien, 2011)

Sadik, Ahmad and Cooper, Tom, *Iraqi Fighters 1953-2003: Camouflage & Markings* (Houston: Harpia, 2008)

Warnes, Alan, *The Pakistan Air Force, 1998 – 2008 A New Dawn* (Stamford: Key Publishing, 2009)

Willisch, Dr. Jürgen, *Mikoyan-Gurevich MiG-21F-13/U* (Arbeitsgemeinschaft Luftwaffe (AGL) e.V., 2004) (in German)

Aviation Archives 03/2008 Vol. 207

Articles from the magazines *FliegerRevue*, *FliegerRevue Extra*, *Air Forces Monthly* and *letectví + kosmonautika*.

Internet Sources (Selection)

http://www.afwing.com/intro/j-7/legend/(abridged Internet version of Aviation Archives 03 / 2008)

https://defence.pk/pdf/threads/mig-21s-f-7s-specifications-capabilities.297826/

https://www.scramble.nl/planning/orbats

https://avialatina.com.ar/los-que-no-fueron-parte-2/

https://www.key.aero/article/history-chengdu-f-7n-iranian-service

https://www.turkishdefencenews.com/bangladesh-supplies-teber-guidance-kit-from-turkiye/

https://www.fullaviacion.com.ar/2022/08/28/rd-93-desde-rusia-con-amor-2-entrega/

https://www.youtube.com/watch?v=BayjAFra_6U

ABOUT THE AUTHOR

Holger Müller

Born in 1968 in the former German Democratic Republic (GDR), Holger Müller was always an aviation enthusiast since his childhood. He completed his military service in the GDR Air Force from 1988 to 1990 as a MiG-21 technician. Since then, this subject has never left him – as evidenced by numerous related articles and books he has published. Müller has travelled to more than 20 user countries of the MiG-21 and J/F-7s fighters and the majority of his publications about this prolific family are based on his own research and photographic material.